D0190822

SNEAKY MATH
A Graphic Primer
with Projects

Other Books in the Sneaky Series

Sneaky Uses for Everyday Things

Sneakier Uses for Everyday Things

Sneakiest Uses for Everyday Things

The Sneaky Book for Boys

The Sneaky Book for Girls

Sneaky Science Tricks

Sneaky Green Uses for Everyday Things

Super Sneaky Uses for Everyday Things

SNEAKY MATH

a graphic primer with projects

Ace the Basics of Algebra, Geometry, Trigonometry, and Calculus with Everyday Things

CY TYMONY

Andrews McMeel Publishing

Kansas City · Sydney · London

CONTENTS

Introduction. vi

Part 1
ARITHMETIC 1

PROJECT Sneaky Cup Calculator. 4
PROJECT Sneaky Napier's Bones Calculator 8
PROJECT Sneaky Fraction Quizzer. 16

Part 2
ALGEBRA 23

PROJECT Sneaky Powers and Square Root Quizzer 29
PROJECT Sneaky Algebra Calculator. 44
PROJECT Algebra Function and Sneaky Temperature Converter 53
PROJECT Coordinate Plane Quizzer. 61

Part 3
GEOMETRY / TRIGONOMETRY 67

PROJECT Euler's Polyhedron Formula Demo 74
PROJECT Sneaky Pi Cups 79
PROJECT Triangle Pythagorean Theorem Demo 82
PROJECT Triangle Pythagorean Theorem Challenge 85
PROJECT Sneaky Hypsometer. 90

Part 4

PRE-CALCULUS 95

PROJECT Calculus vs. Algebra Rate of Change Demo. 100
PROJECT Calculus Differentials and Slope Demonstration 107

Part 5

GOING FURTHER
WITH SNEAKY MATH 113

Scientific Calculators. 114
ALGEBRA Expanding and Simplifying Terms 121
CALCULUS The Power Rule 124
Formula for Determining the Rate of Change 126
Calculating Differentials with a Scientific Calculator. 133
Integral Calculus 136
Calculating Integrals with a Scientific Calculator 140
BONUS PROJECT Integral Math Card 146
Math with Spreadsheets 151
More Creative Math Designs and
    Challenges Using Everyday Things 160

SNEAKY MATH
REFERENCE MATERIAL 163

Modern Math Notables 164
Recommended Books 167
Recommended Websites 169
Glossary 170
Math Conversion Chart. 173

INTRODUCTION

WHAT COLOR WAS YOUR MATH?

Think of your favorite toy or gift from childhood. Remember the first time you saw it in a TV commercial or a store window and how you begged your parents to get it for you? What color was it?

Remember your first science or craft project for school? Most people still do.

Now, what was your favorite memory of math? What was your favorite math project or plaything? How fondly do you remember anything at all about math?

Math is a science, a language, and an art that deals with real and imaginary objects and observations. It is used to count and calculate people, atoms, and celestial bodies. It deals with shapes, measurements, patterns, risk, and rates of change. Almost everything can be spoken of in the language of math.

So why don't your math memories bring a smile to your face? When a subject is thought of only as "drills" and "problems" instead of practical, fun activities and *physical crafts*, it's no wonder it doesn't conjure fond memories!

It's easy to forget things that we see and hear that do not seem practical or relevant, like a math lecture or problem on a blackboard. But we easily remember things that we do like:

- Trips we take
- Activities we participate in
- Items we desire and possess
- Things we make

Traveling is fun. Making something is involving. Doing activities that we relate to and enjoy is memorable. Math should also be memorable. And it can be if we see it in things we can relate to and participate in a physical activity involving it.

Sneaky Math is designed to supplement math educational materials by taking lessons off the written page, computer screen, and blackboard and placing them in your hands. Instead of pages and pages of history and theory that you can't relate to, *Sneaky Math* is a graphics-filled primer with easy-to-make projects designed for maximum clarity and accessibility. It answers the question "What can I do with this right now?" by plunging you into practical activities that you'll want to do and share.

This book tackles the most confusing math symbols and concepts to prepare you to excel in math courses. You'll quickly understand the following topics:

- Division and multiplication with mixed fractions
- Square roots and exponents
- Algebra variables and functions
- Practical formulas used in everyday life
- Geometry and trigonometry techniques
- What calculus is and how it is used
- Unusual math symbols and their usage
- Getting started with scientific calculators
- Using spreadsheets for math operations
- Creating *Sneaky Math* designs and challenges

Sneaky Math assumes the reader has fundamental arithmetic skills. It provides a graphic and DIY theme to make learning math accessible, practical, fun, and most important, memorable. As a bonus, *Sneaky Math* projects and designs are inexpensive and are made to pass along to others.

You don't have to build all the projects that follow; you'll quickly find that a DIY style is more accessible than a typical textbook. You'll quickly understand the topic and undoubtedly fashion a few items to challenge your knowledge and then pass on your devices to others.

Using items found in every household, you'll quickly make Sneaky Calculators and Math Quizzers, and perform experiments you won't easily forget.

Our primary goal is to change your attitude from "that math" into "*my* math."

LET'S GET STARTED!

Part 1

ARITHMETIC

POSITIVE AND NEGATIVE NUMBER RULES

■ **Positive and negative numbers and zero are called integers.**
On a number line, negative numbers are shown at the left of the zero mark and positive numbers to the right of zero.

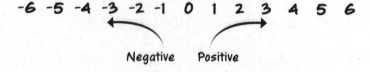

-6 -5 -4 -3 -2 -1 0 1 2 3 4 5 6

Negative   Positive

Here are the rules for adding, subtracting, multiplying, and dividing numbers.

■ **Adding and subtracting positive numbers remain positive:**
$$4 - 2 = 2$$
(except when subtracting a larger number from the first, e.g., 4 – 7)

■ **When adding or subtracting a larger number, the sign of the larger is the sign of the result.**
$$-7 + 1 = -6$$

■ **With multiplication and division between two numbers, Like Signs = a Positive; Unlike Signs = a Negative.**

Like signs = a positive

$-\div-$ or $-\times-$ or $+\times+$ or $+\div+$

ALL EQUAL A POSITIVE NUMBER: No matter how many numbers you are multiplying, an even number of negative (-) numbers gives you a positive result. An odd number of negative (-) numbers gives you a negative answer.

THE RULES

$4 - 2 = 2$ $-7 + 1 = -6$ $8 - -3 = 11$

Double negative—add both numbers as a positive

$-2 - 4 = -6$

Multiplication and Division

Like signs = a positive	Unlike signs = a negative
$-\div-$	$-\div+$
$-\times-$	$-\times+$
$+\times+$	$+\div-$
$+\div+$	$+\times-$

EXAMPLES

$-2 \times 2 = -4$	$-2 \times -3 = 6$
$6 \div -2 = -3$	$4 \times 3 = 12$
$-4 \div 2 = -2$	$6 \div -3 = -2$
$7 \times -4 = -28$	$-8 \div -2 = 4$

Project—
SNEAKY CUP CALCULATOR

You can make a simple adding device using two rulers or paper strips numbered like a ruler taped to two cups.

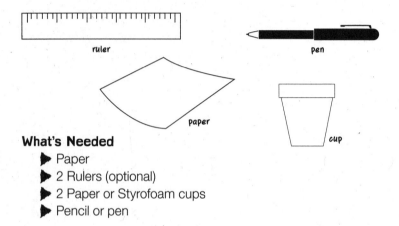

ruler

pen

paper

cup

What's Needed
▶ Paper
▶ 2 Rulers (optional)
▶ 2 Paper or Styrofoam cups
▶ Pencil or pen

As shown in **FIGURE 1**, simply place the rulers or paper strips numbered like a ruler next to each other. Line up a number on the top ruler with the 0 on the bottom one. Following the top number 5, down to the 0, and across to the lower 4 reveals the sum of 9 on the upper ruler. See **FIGURE 2**.

FIGURE 1

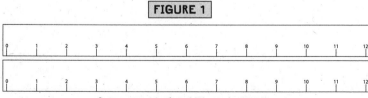

Ruler paper strips for addition and subtraction

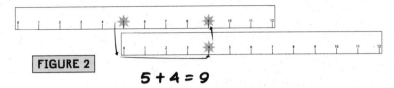

FIGURE 2

$5 + 4 = 9$

Perform the reverse operation for subtraction.

You can copy the ruler strips on paper and tape them at the top of two cups, insert one inside the other, and turn the cups to align the numbers to add and subtract. See **FIGURE 3**.

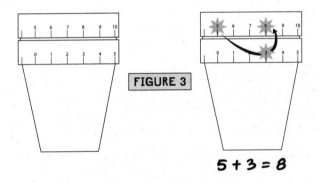

FIGURE 3

$5 + 3 = 8$

GOING FURTHER

Cut out and attach the two multiplication ruler strips to the cups, and now your cup calculator can multiply and divide as shown in **FIGURE 4**. Select your number on the upper strip and line up the 1 on the lower strip with your number. Now go to the number you want to multiply and look to the upper strip for the answer.

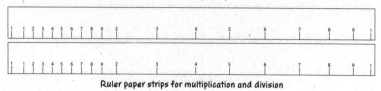

Ruler paper strips for multiplication and division

FIGURE 4

SNEAKY TIP: If the top cup's numbers cannot be seen, place a small piece of crumpled paper in the lower cup to raise the level of the top cup.

No cups available? Make your own with paper and scissors. Just draw ruler "gauges" on two thick pieces of paper. On one ruler gauge, cut two notches at the bottom. See **FIGURE 5**.

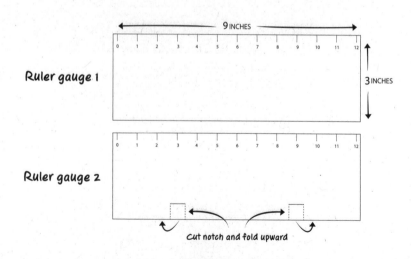

FIGURE 5

Tape each ruler gauge into a loop. On the "cup" with the notches, fold the tabs upward to provide an elevated platform for the upper one. See **FIGURE 6**.

Then, slip the top "cup" into the lower one, and you've got a paper calculator as shown in **FIGURE 7**.

FIGURE 6

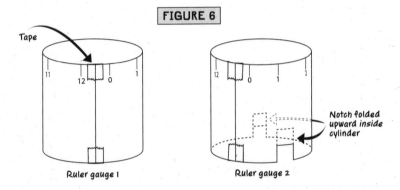

Tape

Ruler gauge 1

Notch folded upward inside cylinder

Ruler gauge 2

FIGURE 7 Start at a number at the top, like 5. With 0 positioned below it, follow the bottom cup over to a number. The number above it will be their sum.

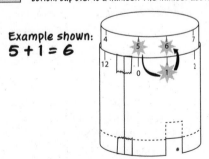

Example shown:
5 + 1 = 6

The lower cylinder's notches hold the upper cylinder in place.

Project—
SNEAKY NAPIER'S BONES CALCULATOR

A Scottish mathematician named John Napier discovered logarithms and invented a simple calculator to multiply any two numbers. The simple calculator is known as Napier's bones because it can be constructed from bone, paper, and so on.

What's Needed
▶ Paper or cardboard
▶ Pen
▶ Scissors

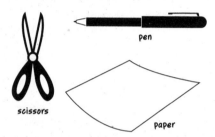

pen

scissors

paper

To make Napier's bones, cut out ten strips of paper, each divided into nine parts. On each strip starting at the second box, a slanted line separates the tens digit from the units digit. Draw or print the strips exactly as shown in **FIGURE 1**. Now you can use the paper strips to multiply two numbers. The No.1 column is shown to the left of the examples to help you envision the multiplier.

To calculate 4812 × 4, place the strips for the numbers being multiplied side by side as shown in **FIGURE 2**. Add the numbers on the slanted rows together, and it will total the sum of 19248.

FIGURE 3 shows how to position the strips to multiply 6375 × 4, totaling 254100.

NOTE: When two numbers on a diagonal line equal ten or more, the tens place is added to the sum of the column to the left. That's why 6375 × 4 = 25500 and not 254100.

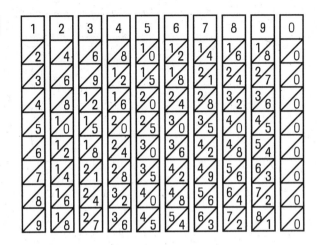

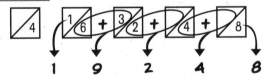

1   9   2   4   8

6375
x 4 = 25500

FRACTIONS

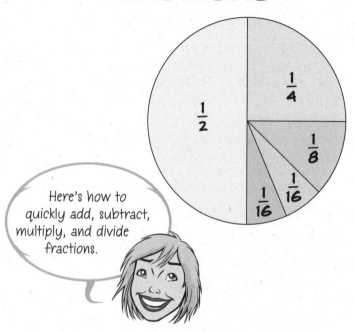

Here's how to quickly add, subtract, multiply, and divide fractions.

- The upper number of a fraction is called the numerator. The lower number is the denominator.

- When fractions have a common denominator, you can simply add the numerators together to find the sum.

- When fractions do not have a common denominator, you must find one. Note: This is only for adding and subtracting.

In the example shown in **FIGURE 1**, to add $\frac{1}{3} + \frac{3}{8} + \frac{4}{12}$ you must find a number their denominators are all multiples of.

FIGURE 1

$$\frac{1}{3} + \frac{3}{8} + \frac{4}{12}$$

Numerator

Denominators are all different.

Finding a Common Denominator

List the multiples of the denominators until a common one is found. Then add the numerators.

FIGURE 2

$$\frac{1}{3}$$

$3 \times 2 = 6$

$3 \times 3 = 9$

$3 \times 4 = 12$

$3 \times 5 = 15$

$3 \times 6 = 18$

$3 \times 7 = 21$

$3 \times 8 = 24$

FIGURE 3

$$\frac{3}{8}$$

$3 \times 3 = 9$

$8 \times 3 = 24$

FIGURE 4

$$\frac{4}{12}$$

$12 \times 2 = 24$

As shown in **FIGURES 2**, **3**, and **4**, the common multiple of 3, 8, and 12 is 24. Multiply each numerator and denominator by the number necessary to bring each denominator to 24, and then add the numerators. For example, for the fraction $\frac{1}{3}$, 24 divided by 3 is 8. Multiply 8 by the numerator, 1, to get 8. So $\frac{1}{3}$ becomes $\frac{8}{24}$.

Similarly, for the fraction $\frac{3}{8}$, 24 divided by 8 is 3. 3 times the numerator, 3, is 9. So $\frac{3}{8}$ is equivalent to $\frac{9}{24}$.

Last, for the fraction $\frac{4}{12}$, 24 divided by 12 is 2. 2 times the numerator, 4, is 8. So $\frac{4}{12}$ is equivalent to $\frac{8}{24}$.

$\frac{8}{24} + \frac{9}{24} + \frac{8}{24}$ equals $\frac{25}{24}$, which equals $1\frac{1}{24}$, as shown in **FIGURE 5**.

FIGURE 5

$$\frac{1}{3} = \frac{8}{24}$$

$$\frac{3}{8} = \frac{9}{24}$$

$$\frac{4}{12} = \frac{8}{24} \qquad 8 + 9 + 8 = \frac{25}{24} \text{ reduced to } 1\frac{1}{24}$$

- To **subtract fractions**, simply subtract the numerators. See **FIGURE 6**.

FIGURE 6 To subtract fractions, focus on just the numerators:

$$\frac{7}{9} - \frac{2}{9} \text{ is just like } 7 - 2, \text{ which equals } 5 \text{ (or } \frac{5}{9})$$

When subtracting a larger fraction from a smaller one, just work with the numerators:

$$\frac{2}{9} - \frac{7}{9} \text{ is just like } 2 - 7, \text{ which equals } -5 \text{ (or } -\frac{5}{9})$$

- **Multiplying fractions** is actually easier. You do not need to find a common denominator. Simply multiply the numerators with each other and do the same with the denominators, as shown in **FIGURE 7**.

FIGURE 7

$$\frac{1}{4} \times \frac{2}{3} = \frac{1 \times 2 = 2}{4 \times 3 = 12} \text{ or } \frac{1}{6}$$

■ You can multiply fractions with a whole number by converting the whole number into a fraction first and then multiplying them. **EXAMPLE:** $\frac{1}{4}$ times $2 = \frac{1}{4}$ times $\frac{2}{1} = \frac{2}{4}$ or $\frac{1}{2}$. See **FIGURE 8** for another example

FIGURE 8 | Multiplying fractions — fraction x whole number

$$\frac{1}{3} \times 4 = \frac{1}{3} \times \frac{4}{1} = \frac{4}{3} \text{ or } 1\frac{1}{3}$$

(four times)

■ To multiply a fraction with a mixed fraction (a number with a fraction), convert the mixed fraction into an improper fraction (its numerator is larger than the denominator) and multiply it by the first fraction. See **FIGURE 9**.

FIGURE 9 | Multiplying fractions — fraction x mixed fraction

$$2\frac{1}{2} \times \frac{2}{3} = \frac{5}{2} \times \frac{2}{3} = \frac{10}{6} = \frac{5}{3} = 1\frac{2}{3}$$

Convert a Mixed Fraction into an Improper Fraction

$$2\frac{1}{2} = 2 \quad \frac{1}{2} \, + \, = \frac{5}{2} = 1\frac{2}{3}$$

$$\left(\frac{2 \times 2 + 1}{2} \right)$$

■ **Dividing fractions** is just about as easy as multiplying them. All you have to do is flip the second fraction and multiply the resulting fractions. See **FIGURES 10** and **11**.

FIGURE 10 | Fraction divided by a whole number

Turn the second fraction upside down and multiply.

$$\frac{1}{3} \div 4 = \frac{1}{3} \div \frac{4}{1} \qquad \frac{1}{3} \times \frac{1}{4} = \frac{1}{12}$$

FIGURE 11   Fraction divided by a fraction

$$\frac{2}{5} \div \frac{1}{6} = \frac{2}{5} \times \frac{6}{1} = \frac{12}{5} = 2\frac{2}{5}$$

Reverse the second fraction and multiply.

■ With two mixed fractions, convert them to improper fractions, flip the second fraction, and multiply the numerators by each other and do the same with the denominators as shown in **FIGURE 12**.

FIGURE 12   Mixed Fractions Divided by Mixed Fractions

$$3\frac{1}{4} \div 2\frac{1}{5} = \frac{13}{4} \div \frac{11}{5} = \frac{13}{4} \times \frac{5}{11} = \frac{65}{44} = 1\frac{5}{11}$$

Reverse order and multiply.

Here's how to work with decimals, percentages, and fractions.

■ In **FIGURE 13** the fraction $\frac{2}{3}$ has a denominator of 3, which cannot divide evenly into 10 or 100, etc., so the number 9 gives you an approximate total. You can multiply by 333 like this: 3 times 333 = 999. 2 times 333 = 666 (or $\frac{6}{9}$, which is the equivalent decimal 0.66666666666667 to be exact).

FIGURE 13

Convert a fraction into a decimal.

$$\frac{2}{3} = 3\overline{)\begin{array}{l} .66 \\ 2.0 \end{array}} = .66 \quad \text{Decimal}$$

$$\begin{array}{r} 18 \\ \hline 20 \end{array}$$

■ To convert the fraction $\frac{5}{8}$ into a percentage (a part of 100), divide 5 by 8 to obtain .625. This represents 62.5 percent, as shown in **FIGURE 14**.

FIGURE 14 Convert a fraction into a percentage.

$$\frac{5}{8} = 8\overline{\smash{)}5.0} \quad = 62.5\%$$

$$\begin{array}{r} .625 \\ 8\overline{)5.0} \\ \underline{48} \\ 20 \\ \underline{16} \\ 40 \end{array}$$

■ To convert the decimal .30 into a fraction, write it as $\frac{30}{100}$, which, when reduced, is $\frac{3}{10}$. See **FIGURE 15**.

FIGURE 15 Convert a decimal into a fraction.

$$.30 = \frac{30}{100}$$

$$\frac{30 \div 10}{100 \div 10} = \frac{3}{10}$$

Project—
SNEAKY FRACTION QUIZZER

Make your own Sneaky Fraction Quizzer to practice your fraction calculating skills. (And later, give it away to a young math student.)

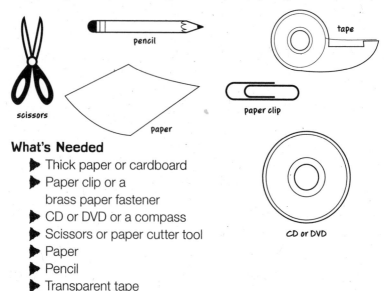

What's Needed
▶ Thick paper or cardboard
▶ Paper clip or a brass paper fastener
▶ CD or DVD or a compass
▶ Scissors or paper cutter tool
▶ Paper
▶ Pencil
▶ Transparent tape

You can use the illustrations on the next pages as a guide or photocopy the pages, paste them on the cardboard, and cut out the shapes shown in **FIGURES 1** through **6**. First, trace a disc the size of a CD or DVD, or use a compass to create a circle with a diameter of $4\frac{3}{4}$ inches, as shown in **FIGURE 1**.

Next, cut out the rectangular cover shown in **FIGURE 2**, including the "window" hole. Or, you can run a pen firmly along the dotted lines to cut away the window section.

Fold the cover in half and puncture a hole in the front as shown in **FIGURE 3**. Also puncture a hole in the center of the disc, as shown in **FIGURE 4**.

Place the disc in the cover and see how the upper one allows you to view the solution when its rectangular opening is properly dialed, as shown in **FIGURE 5**. Carefully push the brass paper fastener or paper clip from the outside of the cover and through the disc.

Dial the disc around to match an equation and view the answer through the window. See **FIGURE 6**.

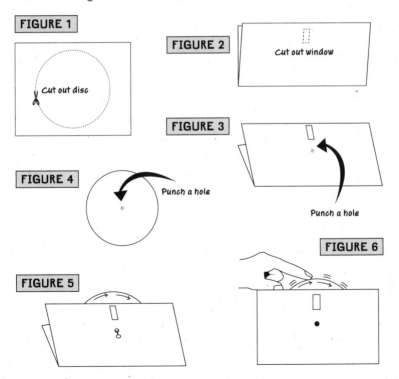

FIGURE 1 — Cut out disc

FIGURE 2 — Cut out window

FIGURE 3 — Punch a hole

FIGURE 4 — Punch a hole

FIGURE 5

FIGURE 6

SNEAKY FRACTIONS

Fractions are part of a whole.

ADDED

MULTIPLIED

DIVIDED

cut along dotted line

$\frac{1}{4}$

$\frac{1}{8}$

$\frac{1}{16}$

$\frac{1}{32}$

$\frac{1}{64}$

$\frac{1}{2}$

Turn the dial and test your fraction skills!

SNEAKY MATH:
A Graphic Primer with Projects

+ ADDING FRACTIONS

$\dfrac{1}{4} + \dfrac{2}{3}$ Find a common denominator.

This is easy! Just multiply the numerators and the denominators.

$$\dfrac{3}{12} + \dfrac{8}{12}$$

$$\dfrac{3}{12} + \dfrac{8}{12} = \dfrac{11}{12}$$

Just add the numerators. You can subtract this way, too!

X MULTIPLYING FRACTIONS

$$\dfrac{3}{4} \times \dfrac{1}{3} = \dfrac{3 \times 1}{4 \times 3} = \dfrac{3}{12} = \dfrac{1}{4}$$

Reduces to

÷ DIVIDING FRACTIONS

Flip

$$\dfrac{1}{5} \div \dfrac{2}{3} = \dfrac{1}{5} \times \dfrac{3}{2} = \dfrac{1 \times 3}{5 \times 2} = \dfrac{3}{10}$$

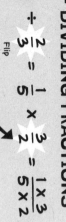

Just flip the second fraction and multiply.

Convert a MIXED FRACTION into an IMPROPER FRACTION

$$3\dfrac{1}{4} = 3\dfrac{1}{4} = \dfrac{13}{4}$$

$$\left(\dfrac{3 \times 4 + 1}{4}\right)$$

Convert an IMPROPER FRACTION into a MIXED FRACTION

$$\dfrac{15}{2} = 2\overline{\smash{\big)}15} = 7\dfrac{1}{2}$$
$$\dfrac{14}{10}$$
$$7.5$$

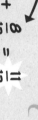

Enlarge on a copier 125%.

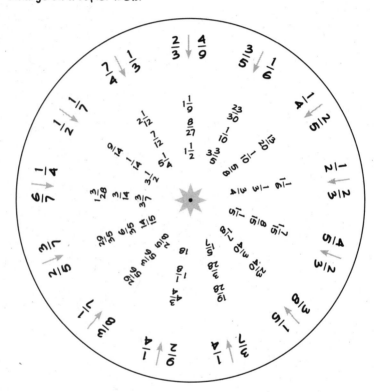

MATH CARD CONSTRUCTION TIPS

In case you do not have (or want to use) a knife or a brass paper fastener, here's how to quickly cut out the window of the math card with scissors and use a paper clip to attach the disc.

You can cut out the window section of the math card with scissors. Start at a side closest to the section to remove, as shown in **FIGURE 1**.

FIGURE 1

To avoid using a knife when cutting a window in the math cards, just use scissors and cut from one side.

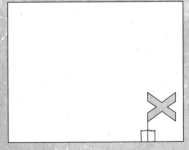

FIGURE 2

Apply transparent tape across the slit.

Then apply a thin layer of transparent tape to the inside of the card to secure the math card. See **FIGURE 2**.

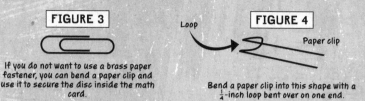

FIGURE 3

If you do not want to use a brass paper fastener, you can bend a paper clip and use it to secure the disc inside the math card.

FIGURE 4

Loop

Paper clip

Bend a paper clip into this shape with a $\frac{1}{4}$-inch loop bent over on one end.

FIGURE 3 shows a paper clip, which, when bent straight with a small loop at one end, can substitute for a paper fastener. See **FIGURE 4**.

Simply push the two ends of the paper clip through the hole on the front of the math card and also through the disc. See **FIGURE 5**.

Secure the two ends to the back of the disc with tape, as shown in **FIGURE 6**.

Now the paper clip allows you to turn the disc around freely. See **FIGURE 7**.

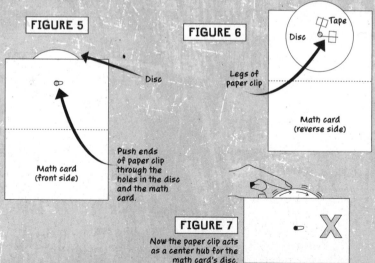

FIGURE 5

Disc

Math card (front side)

Push ends of paper clip through the holes in the disc and the math card.

FIGURE 6

Tape

Disc

Legs of paper clip

Math card (reverse side)

FIGURE 7

Now the paper clip acts as a center hub for the math card's disc.

Part 2

ALGEBRA

MATH SYMBOLS
EXPONENT

An exponent, shown as a superscript at the top right of a number or variable, denotes multiplying the number by itself a certain number of times. If the exponent is 2, this is also called the "power of 2" or a "square."

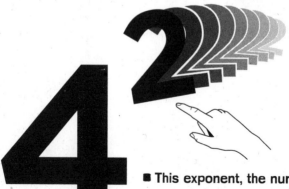

■ This exponent, the number 2, means multiply the number 4 by 4:

$$4^2 = 4 \times 4 = 16.$$
4 times itself 2 times

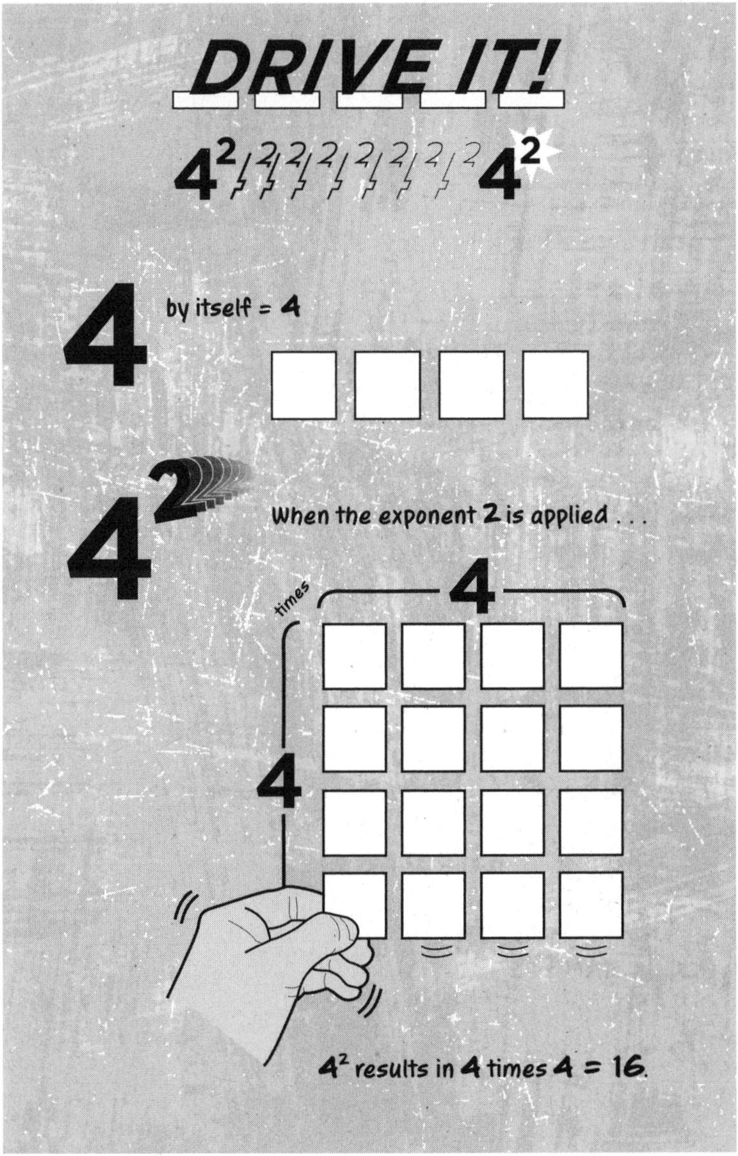

DRIVE IT!

4^2 4^2

4 by itself = **4**

4^2 When the exponent **2** is applied . . .

times

4

4

4^2 results in **4** times **4** = **16**.

4^2

An exponent of **2** is also called "squaring the number":

4   4^2

4 becomes squared to equal 16.

Other exponent amounts can be used, like **3**, which is called the "cube," a number multiplied by itself 3 times.

$4^3 = 4 \times 4 \times 4 = 64$

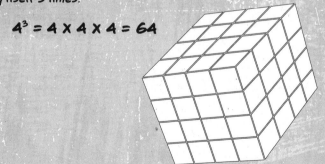

MATH SYMBOLS
SQUARE ROOT

The square root of a number is the opposite of "squaring it."

■ "The square root of 16" means to find what number multiplied by itself equals 16. Answer: 4 also −4. It is like "undoing" a square.

■ The square root symbol is called the radical. The number to be square rooted is called the radicand.

DRIVE IT!

"The square root of **16**" = **4**, because **4 X 4 = 16**.

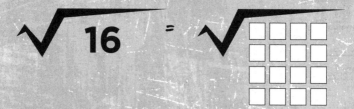

Note: A square root denotes a radical symbol by itself.

It is an abbreviation for the symbol with a small **2** on its top left corner, which means "square" the number.

If we wanted the "cube root" of a number, we would put a **3** in that corner.

What number multiplied times itself **3** times is _____?

Project—
SNEAKY POWERS AND
SQUARE ROOT QUIZZER

Practice your powers and square root calculations with a Sneaky Quizzer you can make with everyday things.

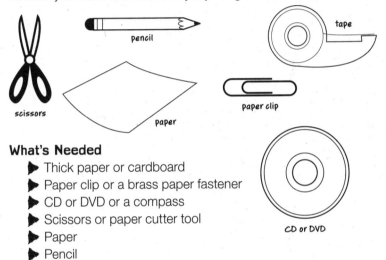

What's Needed
- Thick paper or cardboard
- Paper clip or a brass paper fastener
- CD or DVD or a compass
- Scissors or paper cutter tool
- Paper
- Pencil
- Transparent tape

You can use the illustrations on the next pages as a guide or photocopy the pages, paste them on the cardboard, and cut out the shapes shown in **FIGURES 1** through **6**.

First, trace a disc the size of a CD or DVD, or use a compass to create a circle with a diameter of $4\frac{3}{4}$ inches, as shown in **FIGURE 1**.

Next, cut out the rectangular cover shown in **FIGURE 2** including the "window" holes. Or, you can rub a pen firmly along the dotted lines to cut away the window section.

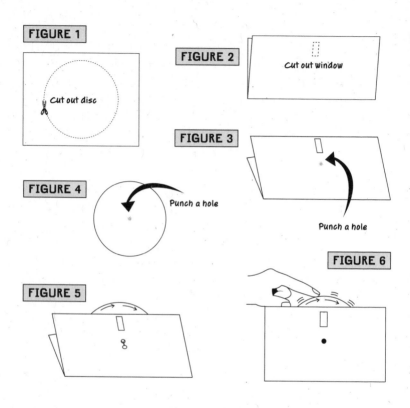

FIGURE 1

Cut out disc

FIGURE 2

Cut out window

FIGURE 3

FIGURE 4

Punch a hole

Punch a hole

FIGURE 6

FIGURE 5

Fold the cover in half and puncture a hole in the front. See **FIGURE 3**. Also puncture a hole in the center of the disc, as shown in **FIGURE 4**.

Place the disc in the cover and carefully push the brass paper fastener or paper clip from the outside of the cover and through the disc as shown in **FIGURE 5**.

Dial the disc around to match an equation, and view the answer through the window. See **FIGURE 6**.

Enlarge on a copier 125%.

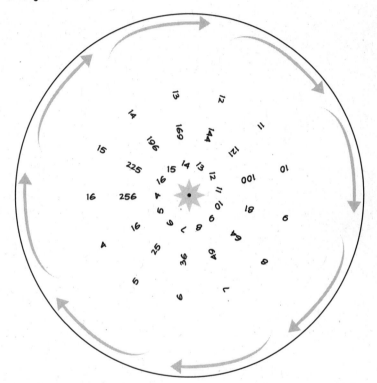

POWERS AND SQUARE ROOTS

POWERS

An exponent, shown at the top right of a number or variable, denotes multiplying the number by itself.

$$4 = \boxed{4}$$

$$4^2 = \boxed{4}$$

$$4^2 = 4 \times 4 = 16$$

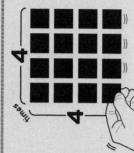

4^2 means multiply 4 by the POWER of 2.
See this graphic EXAMPLE:

SNEAKY MATH:
A Graphic Primer with Projects

POWERS AND SQUARE ROOTS
SQUARE ROOT

Turn the dial and test your powers and square root skills.

Cut along dotted lines.

$$\sqrt{\quad} = 2$$

$$\sqrt{16}$$

The square root of a number is the opposite of "squaring" it.

$$\sqrt{16}$$

means "what number multiplied by itself will equal 16?"

$$\sqrt{16} \Rightarrow \sqrt{16}$$

$$\sqrt{4} \Rightarrow 4 \times 4 = 16$$

MATH SYMBOLS—
VARIABLE

A variable is a stand-in symbol or letter, like X, that is a placeholder for an amount you do not know.

- It is like a package that holds a number—we just do not know what is in the package. But all math rules apply to it.

- Other letters can represent other values. For example, T, V, and D can stand in for time, velocity, and distance.

- EXAMPLE: f(x), or function of x, is a variable that depends on the value of X. Functions define the relationship between inputs and outputs.

DRIVE IT!

Isolate the variable X by getting it alone on one side of an equal sign:

Solve for X:

$$X + 3 = 7$$

$$X + 3 - 3 = 7 - 3$$

These two cancel each other.

$$X + 3 - 3 = 4$$

$$X = 4$$

MATH SYMBOLS—
FUNCTIONS

A function is the relationship between two variables.

$f(x)$

Pronounced "F of X"

■ A function is the relationship between two variables such that for every X there is one and *only* one Y. (Although the variable Y can be the answer to other functions.)

DRIVE IT!

$f(x)$))))))))))) $f(x)$

Formula: $f(x) = X + 5$

MOVIE TICKET	MOVIE TICKET PLUS 3-D FEE

	Movie Ticket Price X	Movie Ticket Price + 3-D Fee $f(x)$
Matinee	$10.00	$15.00
Evening	$15.00	$20.00
VIP Seat	$20.00	$25.00

On a graph

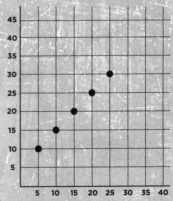

ALGEBRA VARIABLES

Algebra simplifies problems by using variables to represent numbers. (Variables can vary.)

■ **Think of it as arithmetic with letters. A variable can represent a number, a person, or a thing, like the price of a movie ticket, which varies depending on the theater, the time of day, or the type of screening (3-D, IMAX).**

SNEAKY TIP: Since the letter X is frequently used as a variable in algebra, it is not used to represent the multiplication sign. Instead when a variable is next to another letter, or variable, it implies to multiply it times that number or variable. Or a small dot is placed between the number or variable.

EXAMPLE: $4X = 4$ times X.

■ **Equations must balance. Items on each side of the equal sign must have the same amount.**

$1 + 3 = 4$ has the amount, **4**, on each side.

You can determine, or solve, the value of a variable by getting it on one side of the equation.

SNEAKY TIP: To isolate the variable, do the *opposite* operation to the constant number. If a number, like 3, is added to the variable, subtract it from both sides of the equation. If a number is subtracted, simply add it to both sides.

■ **Perform the same operation with multiplication and division.**

■ **Here's how it works:** To solve for X in the equation $X + 2 = 7$, isolate the variable X by removing the constant number 2 from the left side. Since it is added to X, you subtract 2 from both sides to maintain balance, like this:

$$X + 2 = 7 \quad \rightarrow \quad X + 2 - 2 = 7 - 2$$

Since $+ 2 - 2$ cancel each other, you are left with $X = 5$

■ **When variables are on both sides, simply get the variable on one side of the equation like you would a constant number.**

EXAMPLE: $X + 3 = 2X$

$$X + 3 = 2X$$

$$X - X + 3 = 2X - X$$

$$3 = 1X$$

(or $X = 3$)

■ Sneaky Math—Algebra Variables Demo

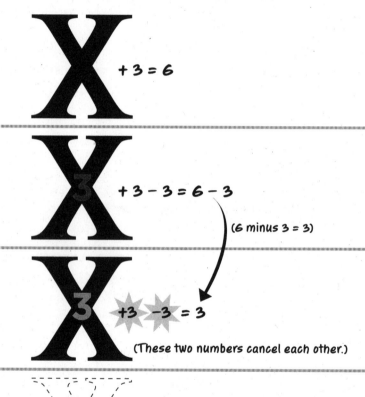

$X + 3 = 6$

$X + 3 - 3 = 6 - 3$

(6 minus 3 = 3)

$X + 3 - 3 = 3$

(These two numbers cancel each other.)

$3 = 3$

EQUATIONS AND FORMULAS

■ **An equation states that two things are equal. It will have an equal sign, like this:**

$$X + 2 = 6$$

What is on the left (X + 2) is equal to what is on the right (6)

■ **A formula is a special type of equation that shows the relationship between different variables.**

EXAMPLE: The formula for finding the volume of a box is:

(Or V = H times W times L)
V = volume, H = height, W = width, and L = length

See **FIGURE 1**.

FIGURE 1

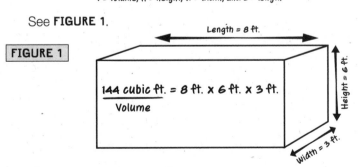

Length = 8 ft.

144 cubic ft. = 8 ft. x 6 ft. x 3 ft.
Volume

Height = 6 ft.

Width = 3 ft.

■ **The Body Mass Index (BMI) formula is:**

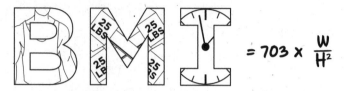 $= 703 \times \dfrac{W}{H^2}$

703 times weight divided by height squared

EXAMPLE: $703 \times \dfrac{140 \text{ lbs.}}{5 \text{ ft. } 6 \text{ inches}^2}$

(or 66 inches x 66 inches)

$\dfrac{98420}{4356}$ = Body Mass Index of **22.59**

■ **The rate formula is:**

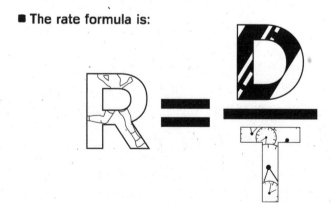

rate = distance divided by time

■ The simple interest formula is:

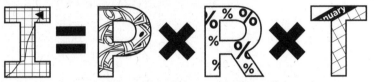

interest = principal times rate times time

EXAMPLE: principal   **$1000.00**   **$1000.00**
                  rate   **5%**   **X .05**
                  time   **1 year**   **X 1**
              interest   **$50.00**

■ The compound interest formula is:

future value = principal times (1 plus rate)
to the power of the number of years

EXAMPLE: principal   **$1000.00**   **$1000.00**
                  rate   **5%**   **X 1.157625**
                  time   **3 years**   **$1157.62**

HERE'S THE MATH: **1000.00 (1 + 0.05)$^{3 \text{ years}}$**
                  principal   rate

ON A CALCULATOR:
        1.05 X 1.05 X 1.05 X 1000 = 1157.625

Project—
SNEAKY ALGEBRA CALCULATOR

Want an easy way to practice algebra variable problems? Then make this Sneaky Algebra Calculator for you or other math students to solve for X wherever you go.

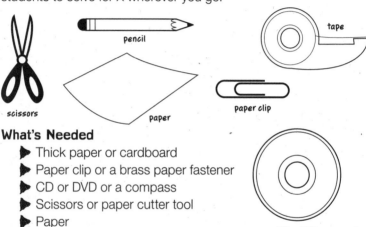

What's Needed
- ▶ Thick paper or cardboard
- ▶ Paper clip or a brass paper fastener
- ▶ CD or DVD or a compass
- ▶ Scissors or paper cutter tool
- ▶ Paper
- ▶ Pencil
- ▶ Transparent tape

You can use the illustrations on the next pages as a guide, or photocopy the pages, paste them on the cardboard, and cut out the shapes shown in Figures 1 through 6.

First, trace two discs the size of a CD or DVD, or use a compass to create a circle with a diameter of $4\frac{3}{4}$ inches, as shown in **FIGURE 1**.

Next, cut out the rectangular cover shown in **FIGURE 2** including the "window" holes. Or, you can rub a pen firmly along the dotted lines to cut away the window section.

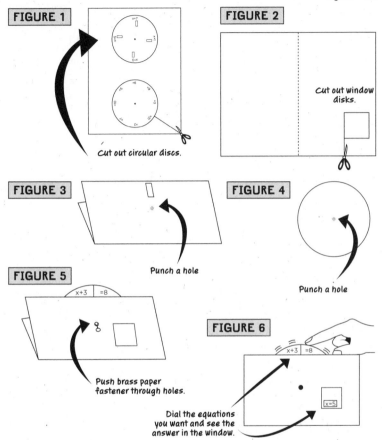

FIGURE 1 Cut out circular discs.

FIGURE 2 Cut out window disks.

FIGURE 3 Punch a hole

FIGURE 4 Punch a hole

FIGURE 5 Push brass paper fastener through holes.

FIGURE 6 Dial the equations you want and see the answer in the window.

Fold the cover in half and puncture a hole in the front, see **FIGURE 3**. Also puncture a hole in the center of the discs, as shown in **FIGURE 4**.

Place the discs in the cover and carefully push the brass paper fastener or paper clip from the outside of the cover and through the disc as shown in **FIGURE 5**.

Dial the discs around and view the information in the window. See **FIGURE 6**.

SIDE 1 Enlarge on a copier 125%.

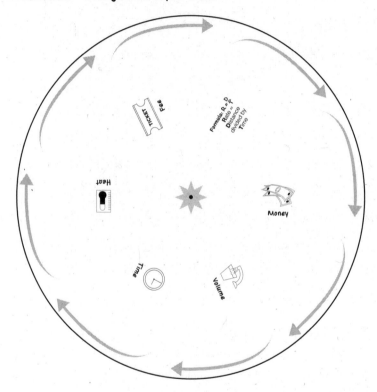

SIDE 2 **Enlarge on a copier 125%.**

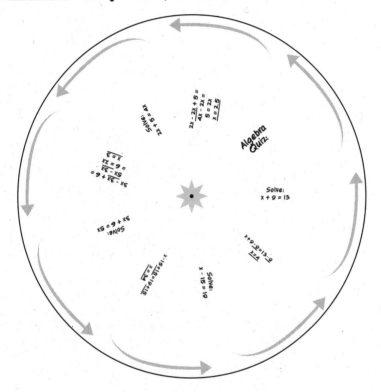

ALGEBRA VARIABLES

A variable, like the letter X, can represent:

cut along dotted line

A variable can be a number, a person, or a thing, like a movie ticket.

Variables, like X or C or F, allow you to make equations and time-saving formulas.

A variable is a placeholder for an amount that you do not know.

A dependent variable, or function, like f(x), depends on the value of X.

SNEAKY MATH:
A Graphic Primer with Projects

ALGEBRA VARIABLES

Equations must balance.
Items on each side of the equal sign
must have the same amount.

1 + 3 = 4

has the amount of **4** on each side.

To solve for X in the equation X + 3 = 6,
remove the constant number 3 from both sides.

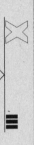

Remove 3 on left side.

Remove 3 on right side.

The equation balances: X = 3

You can solve for the value of a variable by getting it on one side of the equation.

cut along
dotted line

When variables, like X,
are on both sides of an
equation, isolate them
in this manner.

X + 3 = 2X → X − X + 3 = 2X − X

Cancel each other

X − X = 3 = 1X or X = 3

GOING FURTHER

USING ALGEBRA IN WORD PROBLEMS

■ **Allison is three years older than Jennifer. Jennifer is 22 years old. How old is Allison?**

The variable A will represent Allison. 22 represents the age of Jennifer, so the equation looks like this:

$$A - 3 = 22$$

Let's solve for A: $A - 3 + 3 = 22 + 3$

$$A = 25$$

You can use this procedure to determine the time to complete a goal.

EXAMPLE 1: Cliff can carve 5 arrows per hour. How long will it take Cliff to make 15 arrows?

Formula: $\dfrac{G}{R} = T$

Goal / Rate = Time

(goal divided by the work rate = time to complete)

$$\frac{15}{5} = 3$$

(15 arrows divided by 5 Cliff makes per hour = 3 hours to complete)

(Use the same procedure when there are two or more people involved.)

EXAMPLE 2: Cliff can carve 5 arrows per hour. Ty can only carve 3 arrows per hour. If they work together, how long will it take them both to carve 15 arrows?

$$\frac{15}{8} = 1.8$$

(Cliff's 5 arrows + Ty's 3 arrows equals 8 per hour; 15 arrows divided by 8 = 1.8 hours to complete)

EXAMPLE 3: Carlos can clean an apartment in 6 hours. Charles can clean it in 4 hours. Working together, how long will it take them to complete the job?

Using the formula $\frac{G}{R} = T$, we have Goal = 1, or $\frac{12}{12}$ (when the entire apartment is cleaned), Rate = $\frac{5}{12}$ of the Goal per hour ($\frac{3}{12} + \frac{2}{12} = \frac{5}{12}$), Time = $\frac{5}{12}$ = 2.4-hours to completely clean apartment.

Carlos's rate is $\frac{1}{6}$ of the job per hour. Charles's rate is $\frac{1}{4}$ per hour. Finding common denominators and changing these fractions to $\frac{2}{12}$ and $\frac{3}{12}$, their sum is $\frac{5}{12}$ per hour. So, 12 divided by 5 is 2.4, and it will take 2 hours and 24 minutes to clean the house.

TRAVEL RATE WORD PROBLEM

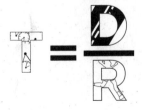

■ 2 superheroes are 300 miles apart. Super Bill flies at 60 miles per hour while Action Annie soars at 40 mph. If they fly toward each other at full speed, how long will it take until they meet?

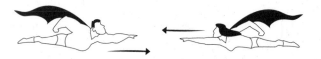

Using the formula **distance / rate = time:**

300 miles (apart) / 60 mph + 40 mph = 100 mph.
(60 mph + 40 mph are added because their closing rate is 100 mph.)

300 miles / 100 mph equals 3 hours until they meet.

WORK PROBLEM

■ **Bill can paint a house in 4 days. Kevin can do it in 5 days. Peter can paint one in 2 days. How long will it take for the house to be painted if they all work together?**

First, calculate how much of the job each person completes per day.

Bill: $\frac{1}{4}$ of the job

Kevin: $\frac{1}{5}$ of the job

Peter: $\frac{1}{2}$ of the job.

Next, find a common denominator for all 3 fractions (20). Then convert and add the fractions:

$$\frac{5}{20} + \frac{4}{20} + \frac{10}{20} = \frac{19}{20}$$

So working together, all 3 men can paint the house in slightly less than 23 hours—22.8 hours to be exact, as shown in the steps below.

Let's cross multiply $\frac{19}{20}$ times $\frac{x}{24}$ to find the rate:

X times 20 = 20X. 19 times 24 = 456.

So 20X = 456. 456 divided by 20 = 22.8 hours

Project—
ALGEBRA FUNCTION AND SNEAKY TEMPERATURE CONVERTER

Practice your algebra function calculations with a Sneaky Quizzer you can make with everyday things.

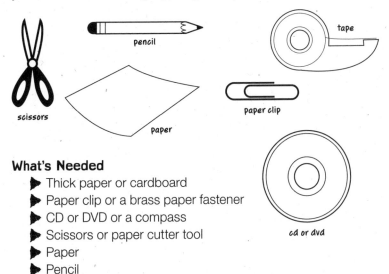

What's Needed
- Thick paper or cardboard
- Paper clip or a brass paper fastener
- CD or DVD or a compass
- Scissors or paper cutter tool
- Paper
- Pencil
- Transparent tape

You can use the illustrations on the next pages as a guide or photocopy the pages, paste them on the cardboard, and cut out the shapes shown in **FIGURES 1** through **6**.

Trace a disc the size of a CD or DVD, or use a compass to create a circle with a diameter of $4\frac{3}{4}$ inches, as shown in **FIGURE 1**.

Next, cut out the rectangular cover shown in **FIGURE 2** including the "window" holes.

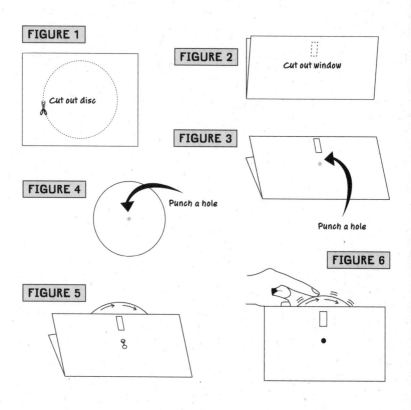

FIGURE 1

Cut out disc

FIGURE 2

Cut out window

FIGURE 3

Punch a hole

FIGURE 4

Punch a hole

FIGURE 5

FIGURE 6

Fold the cover in half and puncture a hole in the front as shown in **FIGURE 3**. Also puncture a hole in the center of the disc, as shown in **FIGURE 4**.

Place the disc in the cover and carefully push the brass paper fastener or paper clip from the outside of the cover and through the disc, as shown in **FIGURE 5**.

Enlarge on a copier 125%.

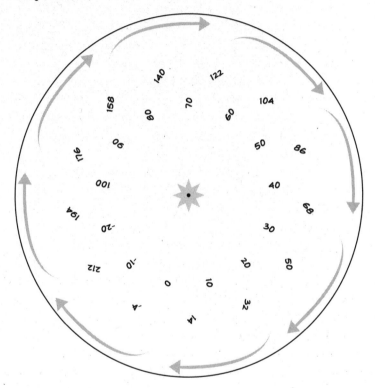

Dial the disc to match an equation and view the answer through the window. See **FIGURE 6**.

Review the algebra function primer and temperature converter information on the front and back of the card and then dial the discs around to view the quiz questions and answers.

SNEAKY TEMPERATURE CONVERTER

FAHRENHEIT

Cut along dotted lines.

CELSIUS

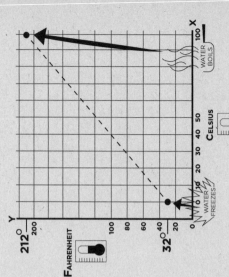

The algebra formula below converts a Celsius temperature to Fahrenheit and vice versa.

$X - 32° \ \mathbf{F} \times \frac{5}{9} = \mathbf{C}$

Function: $f(x) = (X-32) \times \frac{5}{9}$

EXAMPLE: $75° - 32 = 43 \times \dfrac{5}{9} = \dfrac{22}{8} = 23.8°$ Celsius

Fahrenheit

$\mathbf{C} \times 1.8 + 32° = \mathbf{F}$

Function: $f(x) = X \times 1.8 + 32°$

EXAMPLE: $10° \times 1.8 = 18 + 32° = 50°$ Fahrenheit

Celsius

ALGEBRA FUNCTIONS

Think of purchasing a product called X.
The product plus 10% tax is a function.

EXAMPLE: $f(x) = X + .1X$

X $f(x)$

MOVIE
TICKET

MOVIE
TICKET
10% 3D FEE

A function is the relationship between two variables. A function is a predefined formula, which uses the $f(x)$ notation. It means function of X.

Adding 2 to a number is a function.

Example 1: $f(x) = X + 2$

X		$f(x)$
If X = 1	then	$f(x) = 3$
If X = 2	then	$f(x) = 4$
If X = 3	then	$f(x) = 5$

Example 2: $f(x) = X^2$

X		$f(x)$
If X = 2	then	$f(x) = 4$
If X = 3	then	$f(x) = 9$
If X = 4	then	$f(x) = 16$
If X = 5	then	$f(x) = 25$
If X = 6	then	$f(x) = 36$

In this example the variable $f(x)$ equals the value of variable $X + 2$. Algebra functions can be displayed on a graph to view trends, as shown in Example 1.

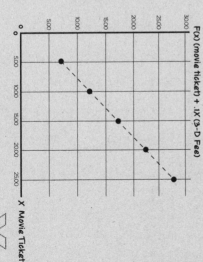

F(x) (movie ticket) + .1X (3-D Fee)

X Movie Ticket

ALGEBRA

COORDINATE PLANE

■ A coordinate plane has numbered or lettered columns and rows displayed on a graph. Examples include a vending machine, a map, a chessboard, and the Battleship game.

■ A coordinate plane allows you to gain visual insights into an algebraic formula or equation.

The vertical plane or axis represents the variable Y, also known as the range.

The horizontal plane or axis represents the variable X, also known as the domain.

■ When two or more points are shown on a coordinate plane, a slope is produced. The steepness or slope of a line can represent an actual slope angle or a rate of time, distance, and more.

Slope is also known as vertical change / horizontal change.

■ **FIGURE** 1 shows a single point on a graph. It is located at (3,4), 3 lines across on the X-axis and 4 up on the Y-axis.

The two points on the graph in **FIGURE 2** are located at (1,1) and (3,3).

To find the slope of the line they form, use the slope formula:

$$Slope = \frac{RISE}{RUN} \quad or \quad \frac{Y2 - Y1}{X2 - X1}$$

The math for finding the slope of the coordinate points is displayed on **FIGURE 3**.

FIGURE 1 Cartesian coordinate plane

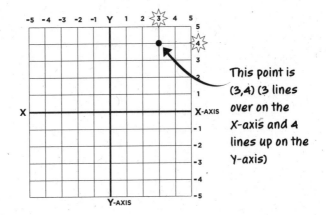

This point is (3,4) (3 lines over on the X-axis and 4 lines up on the Y-axis)

FIGURE 2 Slope of a line (on the coordinate plane)

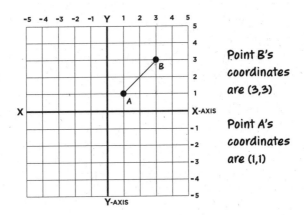

Point B's coordinates are (3,3)

Point A's coordinates are (1,1)

FIGURE 3 The formula for determining the slope of a line is:

$$\text{Slope} = \frac{y_2 - y_1}{x_2 - x_1}$$

This is also known as the rise over run.

Or slope = rise/run, or the change in Y divided by the change in X.

Using the line coordinates above, (1,1) (3,3):
Slope= (3–1) / (3–1) = 2/2 or 1

$$\text{Slope} = \frac{3-1}{3-1} = \frac{2}{2} = 1$$

Project—
COORDINATE PLANE QUIZZER

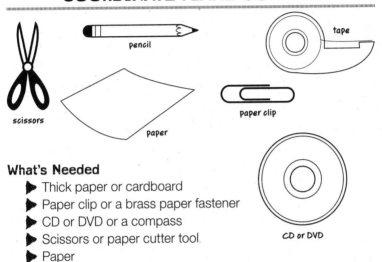

scissors

pencil

tape

paper

paper clip

CD or DVD

What's Needed
- Thick paper or cardboard
- Paper clip or a brass paper fastener
- CD or DVD or a compass
- Scissors or paper cutter tool
- Paper
- Pencil
- Transparent tape

You can use the illustrations on the next pages as a guide or photocopy the pages, paste them on the cardboard, and cut out the shapes shown in **FIGURES 1** through **6**.

First, trace a disc the size of a CD or DVD, or use a compass to create a circle with a diameter of $4\frac{3}{4}$ inches, as shown in **FIGURE 1**.

Next, cut out the rectangular cover shown in **FIGURE 2** including the "window" hole.

Fold the cover in half and puncture a hole in the front, see **FIGURE 3**. Also puncture a hole in the center of the disc, as shown in **FIGURE 4**.

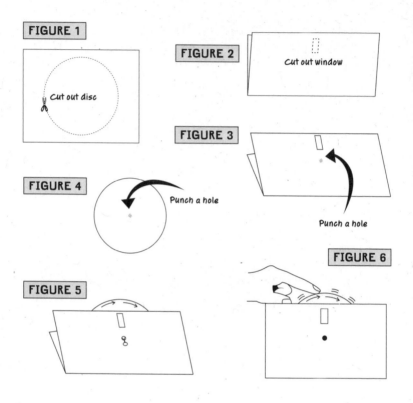

FIGURE 1 — Cut out disc

FIGURE 2 — Cut out window

FIGURE 3 — Punch a hole

FIGURE 4 — Punch a hole

FIGURE 5

FIGURE 6

Place the disc in the cover and carefully push the brass paper fastener or paper clip from the outside of the cover and through the disc.

Dial the disc to see the slope formula and example. See **FIGURE 5**.

Review the coordinate plane primer information on the front and back of the card and then dial the discs around to view the quiz questions and answers. See **FIGURE 6**.

Enlarge on a copier 125%.

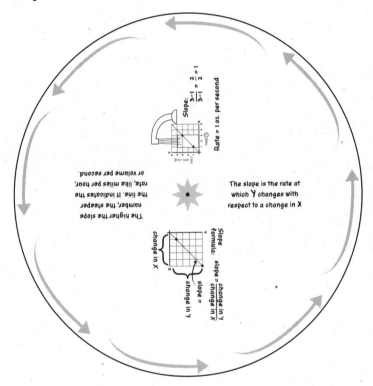

Slope: $\frac{3-1}{3-1} = \frac{2}{2} = 1$

Rate = 1 oz. per second

The higher the slope number, the steeper the line. It indicates the rate, like miles per hour, or volume per second.

The slope is the rate at which Y changes with respect to a change in X

Slope formula: slope = $\frac{change\ in\ Y}{change\ in\ X}$

slope = change in Y

change in X

COORDINATE PLANE QUIZZER

Coordinate planes allow you to gain visual insights of an algebra formula or equation.

cut along dotted line

The vertical plane or axis is represented by the variable Y. The horizontal plane or axis is represented by the variable X.

Slope formula: $slope = \dfrac{change\ in\ Y}{change\ in\ X}$

Slope of line

Point (8,8)

Point (7,6)

The line's slope is:
$\dfrac{8-6}{8-7} = \dfrac{2}{1} = 2$

This is point (2,1) 2 across on X, 1 up on Y

Volume   Rate   Velocity

The steepness or slope of a line can represent an actual slope angle or a rate of time, distance, and more.

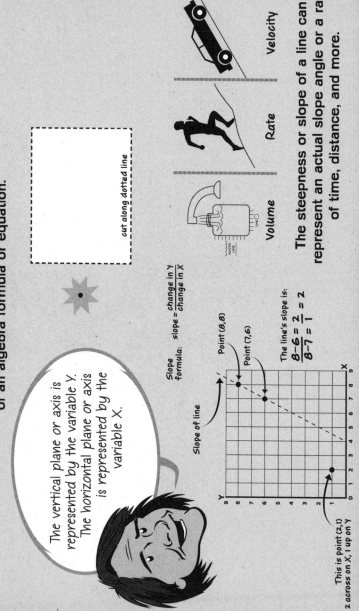

COORDINATE PLANE QUIZZER

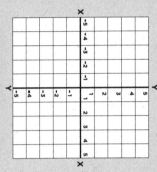

Coordinate Plane

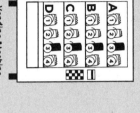

Vending Machine

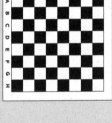

Chessboard

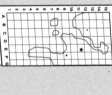

Map

Battleship Game

A coordinate plane has numbered or lettered columns and rows displayed on a graph. Examples include a vending machine, a chessboard, a map, and the Battleship game.

ALGEBRA

GOING FURTHER

ORDER OF OPERATIONS

- When a problem has numbers inside brackets and exponents/powers and square roots, you must follow a procedure to calculate parts of the problem in a particular order. You can remember this order with this acronym:

BODMAS

First, calculate items inside **BRACKETS** first. Then work out the **ORDERS** (powers and square roots) or any grouping symbol.

Next, perform **DIVISION** and **MULTIPLICATION** (left to right), and last, **ADDITION** and **SUBTRACTION** (left to right).

EXAMPLE: To calculate $4 + 6 \times 2$, multiplication before addition: first $6 \times 2 = 12$, then $4 + 12 = 16$.

EXAMPLE: To calculate $(1 + 6) \times 2$, brackets first: first $(1 + 6) = 7$, then $(7) \times 2 = 14$.

Part 3:

GEOMETRY

TRIGONOMETRY

MATH SYMBOLS
THETA

The theta symbol represents an angle of interest.

DRIVE IT!

When you do not know what an angle is . . .

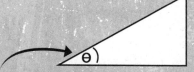

Use the theta symbol to represent the angle of interest in a math problem.

MATH SYMBOLS
PI

π

represents the number 3.14156535 . . . (it goes on forever!). Pi is the ratio of the circumference of a circle to its diameter.

DRIVE IT!

If you know the diameter of a circle, you can use pi to find the circumference.

$$\pi = \frac{C}{D}$$

CIRCLE FORMULAS: $2R = D$
(2 times the radius = the diameter)

$$2R\pi = D\pi$$
(2 times the radius times pi = the diameter times pi)

$$C = D\pi$$
(circumference = the diameter times pi)

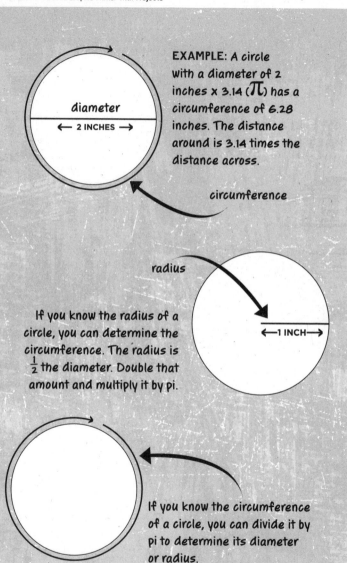

diameter

← 2 INCHES →

EXAMPLE: A circle with a diameter of 2 inches x 3.14 (π) has a circumference of 6.28 inches. The distance around is 3.14 times the distance across.

circumference

radius

←1 INCH→

If you know the radius of a circle, you can determine the circumference. The radius is $\frac{1}{2}$ the diameter. Double that amount and multiply it by pi.

If you know the circumference of a circle, you can divide it by pi to determine its diameter or radius.

GEOMETRY PRIMER

■ **Geometry can be divided into:** EUCLIDEAN GEOMETRY, which includes flat shapes like lines, circles, and triangles—shapes that can be drawn on a piece of paper. **SOLID GEOMETRY** concerns three-dimensional objects like cubes, prisms, and pyramids.

Geometry is all about lines, shapes, angles, space, and their properties.

■ **Geometric Shapes and Formulas**

FIGURE 1 shows the names and number of sides of common polygon shapes:

Triangle, 3 sides   Quadrilateral, 4 sides or angles   Pentagon, 5 sides

Hexagon, 6 sides   Heptagon, 7 sides   Octagon, 8 sides

Nonagon, 9 sides   Decagon, 10 sides   Hendecagon, 11 sides

Dodecagon, 12 sides   Penatadecagon, 15 sides   Icosagon, 20 sides

■ Shape Area Formulas

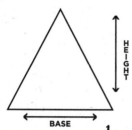

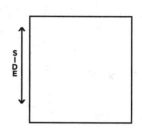

TRIANGLE: *area* $= \frac{1}{2}$ BH
(BH=base x height)

SQUARE: *area* = side^2
(length of 1 side squared)

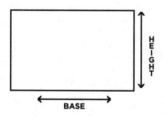

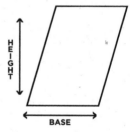

RECTANGLE: *area* = BH

PARALLELOGRAM OR RHOMBUS: *area* = BH

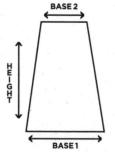

TRAPEZOID:
area $= \frac{1}{2}$ H(B1 + B2)
(area = [Base 1 + Base 2] x vertical Height x $\frac{1}{2}$)

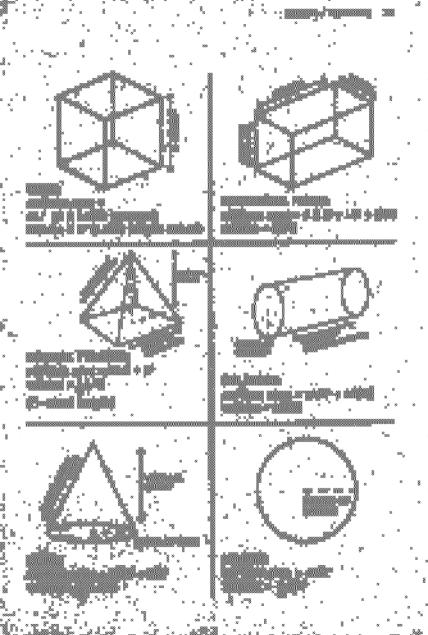

CUBE:
surface area =
6L² (6 x 1 side squared)
volume = L³ (1 side length cubed)

RECTANGLE PRISM:
surface area = 2 (LH + LW + HW)
volume = LWH

SQUARE PYRAMID:
surface area = 2LS + L²
volume = $\frac{1}{3}$ L²H
(S=slant length)

CYLINDER:
surface area = 2πR² + 2πRH
volume = πR²H

CONE:
surface area = πRS + πR²
volume = $\frac{1}{3}$ π R²H

SPHERE:
surface area = 4πR²
Volume = $\frac{4}{3}$ πR³

Project—
EULER'S POLYHEDRON FORMULA DEMO

Mathematician Leonhard Euler discovered a common trait of flat-edged geometric shapes without curves: They follow this formula:

$$V - E + F = 2$$

V = vertex,
which is the point where multiple sides meet

E = edge,
which is the area where 2 sides meet

F = face,
which is the surface area of a side

You can demonstrate Euler's polyhedron formula with a block of cheese.

knife

cheese block

What's Needed
- Cheese block
- Knife

What to Do
FIGURE 1 displays the definition of variables V, E, and F.

FIGURE 1

**Euler's polyhedron formula
(for convex shapes)**

$$V - E + F = 2$$

**Number of vertices minus number of edges
plus number of faces equals 2.**

See **FIGURE 2**.

No matter the shape of a polyhedron, the formula always results in the number 2 (except with special computer-generated convex shapes). See **FIGURE 3**.

FIGURE 4 shows a 4-sided tetrahedron with 4 vertices, 6 edges, and 4 faces. This results in

$$4 - 6 + 4 = 2$$

FIGURES 5, **6**, and **7** show similar instances where different shapes of polyhedrons adhere to Euler's polyhedron formula.

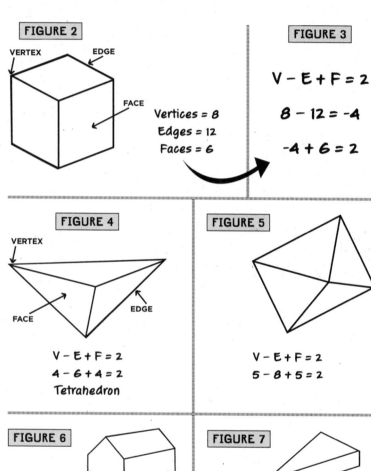

FIGURE 2

VERTEX　EDGE

FACE

Vertices = 8
Edges = 12
Faces = 6

FIGURE 3

$V - E + F = 2$

$8 - 12 = -4$

$-4 + 6 = 2$

FIGURE 4

VERTEX

FACE　EDGE

$V - E + F = 2$
$4 - 6 + 4 = 2$
Tetrahedron

FIGURE 5

$V - E + F = 2$
$5 - 8 + 5 = 2$

FIGURE 6

$V - E + F = 2$
$10 - 15 + 7 = 2$

FIGURE 7

$V - E + F = 2$
$6 - 9 + 5 = 2$
$-3 + 5 = 2$

To physically prove this, start with a square or rectangular block of cheese. See **FIGURE 8**. Cut off one end at a diagonal angle. See **FIGURE 9**.

Euler's Polyhedron Formula Demo

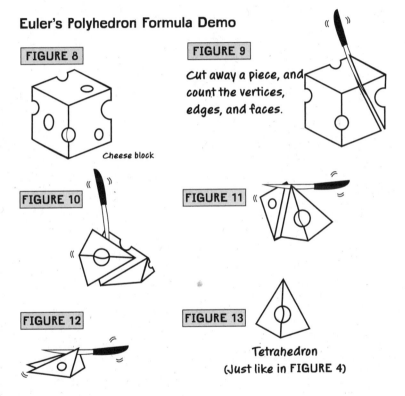

FIGURE 8

Cheese block

FIGURE 9

Cut away a piece, and count the vertices, edges, and faces.

FIGURE 10

FIGURE 11

FIGURE 12

FIGURE 13

Tetrahedron
(Just like in FIGURE 4)

Count the vertices, edges, and faces that remain. Use Euler's formula and write down your results.

Keep cutting away straight pieces from the block of cheese, and you'll find that all shapes will adhere to the formula. See **FIGURES 10**, **11**, **12**, and **13**.

CIRCLES—DIAMETER AND CIRCUMFERENCE

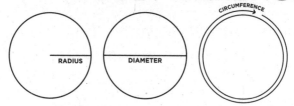

- The radius of a circle is the distance from the center to the edge. The diameter starts at one side of the circle, goes through the center, and ends on the other side.

- The circumference is the distance around the edge of the circle.

- When you divide the circumference by the diameter you get 3.1415 . . . , which is the number pi.

$$\pi = 3.14$$

Circle Formulas:

Circumference = π D
or
C = 2 π R
or
C = π D
or
C = 2 R π

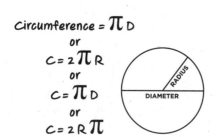

Area = πR^2
(area = pi times radius squared)

Project—
SNEAKY PI CUPS

Make and give away pi cups, which spread the word about the classic formula of a circle.

What's Needed
▶ Paper or Styrofoam cups
▶ Marker pen

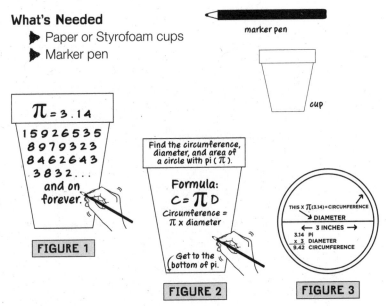

FIGURE 1

Find the circumference, diameter, and area of a circle with pi (π).

Formula:
C = π D
Circumference = π x diameter

Get to the bottom of pi.

FIGURE 2

FIGURE 3

Make It

Simply write the symbol and number for pi on one side of a cup, as shown in **FIGURE 1**.

On the other side, write the definition of pi with a math challenge, as shown in **FIGURE 2**.

On the bottom of the cup, measure its diameter and draw the information shown in **FIGURE 3** to show how to calculate the cup's circumference.

TRIGONOMETRY PRIMER

A triangle is the simplest shape since it has the smallest number of straight lines.

- The three angles in a triangle always add up to 180°.

- Triangle types:

Equilateral Triangle
Three equal sides
Three equal angles
(always 60°)

Acute Triangle
All angles are less than 90°.

Isosceles Triangle
Two equal sides
Two equal angles

Right Triangle
Has a right angle (90°)

Obtuse Triangle
Has an angle more than 90°

Scalene Triangle
No equal sides
No equal angles

A **reflex angle** is
90° to 360°

- Triangle Area Formula

$$\text{Area} = \tfrac{1}{2} \times B \times H$$

(The area is half of the base times the height.)
- B is the distance along the base.
- H is the height (measured at right angles to the base).
- A triangle is always $\tfrac{1}{2}$ the area of a rectangle with the same base and height.

PYTHAGOREAN THEOREM

The Pythagorean theorem states that the square of the longest side of a right-angle triangle (the side opposite the right angle) is equal to the sum of the squares of the other two sides.

The formula looks like this: $A^2 + B^2 = C^2$. See **FIGURE 1**.

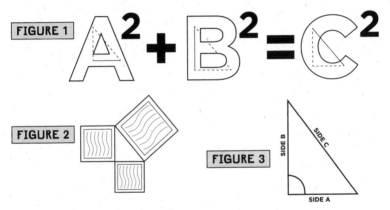

FIGURE 1

FIGURE 2

FIGURE 3

SIDE B

SIDE C

SIDE A

FIGURES 2 and **3** show the physical version of "squaring" sides A, B, and C.

In the triangle shown in **FIGURE 4**, Side A is 3 inches and Side B is 4 inches.

Side A squared is 9, Side B squared is 16. Their sum is 25. The square root of 25 is 5.

Side C is 5 inches long.

With the Pythagorean theorem, you can calculate shortcuts to a path, as shown in **FIGURE 5**.

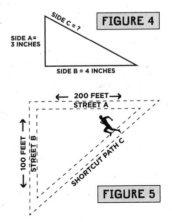

SIDE C = ?
FIGURE 4
SIDE A = 3 INCHES
SIDE B = 4 INCHES

← 200 FEET →
STREET A
100 FEET
STREET B
SHORTCUT PATH C
FIGURE 5

Project—
TRIANGLE PYTHAGOREAN THEOREM DEMO

It's easy to prove the Pythagorean theorem ($A^2 + B^2 = C^2$) with a hands-on demonstration you'll never forget using ordinary paper or cardboard.

What's Needed
- Paper or cardboard
- Scissors
- Pen

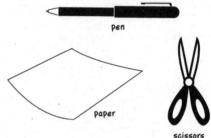

pen

paper

scissors

Make It
Draw a right-angle triangle as shown in **FIGURE 1**. Label the sides as shown in **FIGURE 2**.

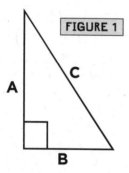

FIGURE 1

C

A

B

Pythagorean Theorem
(concerns right-angle triangles)
Side A squared + Side B squared =
Side C squared

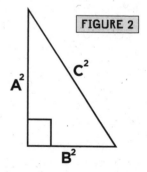

FIGURE 2

C^2

A^2

B^2

Squares of the sides of triangle
The area of A^2 and B^2 fits in C^2. If the length of
A = 3-inches, then A^2 (A squared) = 3-inches
times 3-inches = 9-inches.
So A "squared" = 9-inches.

MATH FORMULAS ILLUSTRATED

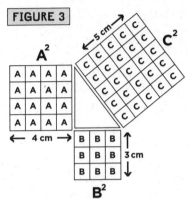

FIGURE 3

Cut out and label three square shapes that are equal to the three sides of the triangle, as shown in **FIGURE 3**. Position them near each side of the triangle to show how they fit.

Note: Side A and Side B are the shortest sides of a right-angle triangle.

Side C, opposite the right angle, is the longest side of a right-angle triangle.

Next, cut out two rectangle pieces of Side A, as shown in **FIGURE 4**.

FIGURE 4

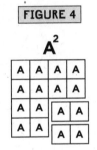

Example 1: Side A = 4 inches
Side B = 3 inches
What is the length of side C?
$A^2 = 3 \times 3 = 9$
$B^2 = 4 \times 4 = 16$
$A^2 + B^2 = 25$
Square root of 25 = 5
Side C = 5 feet because
$C^2 = 5^2 = 25$

FIGURE 5

Remove the square from Side C, and place the large piece of Side A against the triangle's Side C. See **FIGURE 5**.

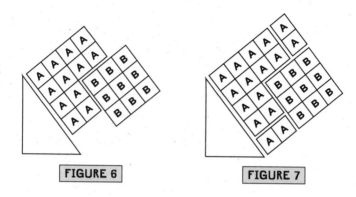

FIGURE 6   **FIGURE 7**

Position the square of Side B against the Side A square, as shown in **FIGURE 6**.

Last, place the two small pieces cut from Side A into position to complete the square. See **FIGURE 7**.

As you can see, the square shapes of A^2 and B^2 equal the space of C^2.

Project—
TRIANGLE PYTHAGOREAN THEOREM CHALLENGE

Make a simple yet challenging device with paper clips and straws to teach the Pythagorean theorem.

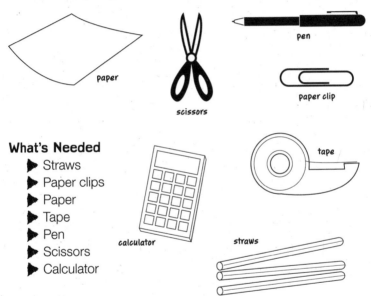

paper

scissors

pen

paper clip

What's Needed
- Straws
- Paper clips
- Paper
- Tape
- Pen
- Scissors
- Calculator

tape

calculator

straws

Make It

Bend three paper clips into right-angle and acute-angle (less than 90-degree angle) shapes as shown in **FIGURE 1**.

FIGURE 1

Bend paper clips into various angles.

FIGURE 2

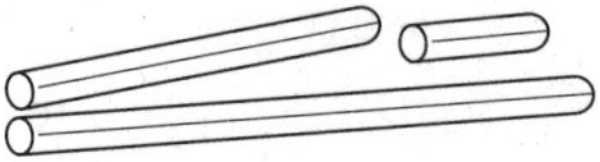

Cut two straws into various lengths.

FIGURE 3

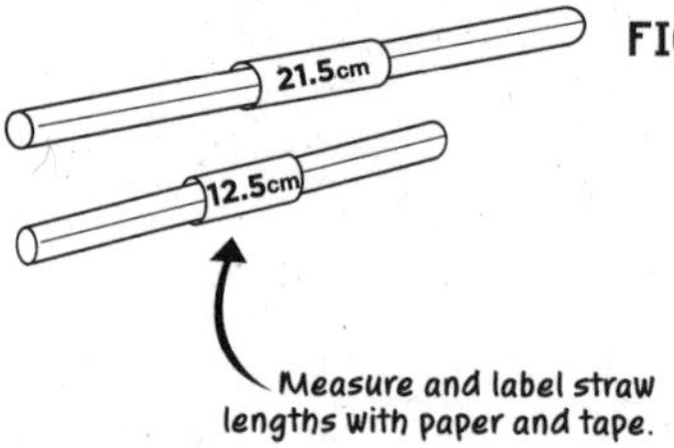

21.5cm

12.5cm

Measure and label straw lengths with paper and tape.

FIGURE 4

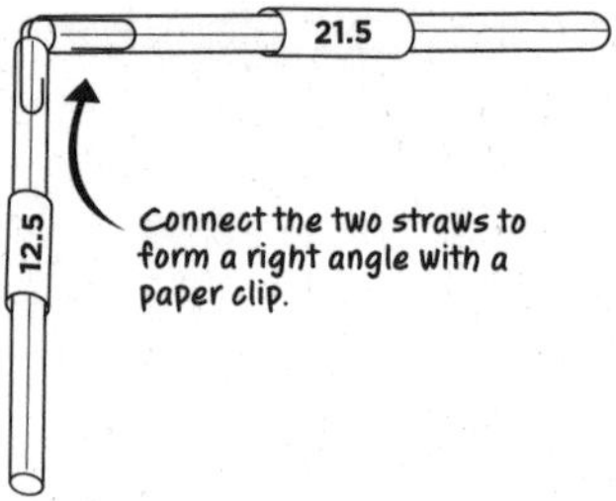

21.5

12.5

Connect the two straws to form a right angle with a paper clip.

FIGURE 5

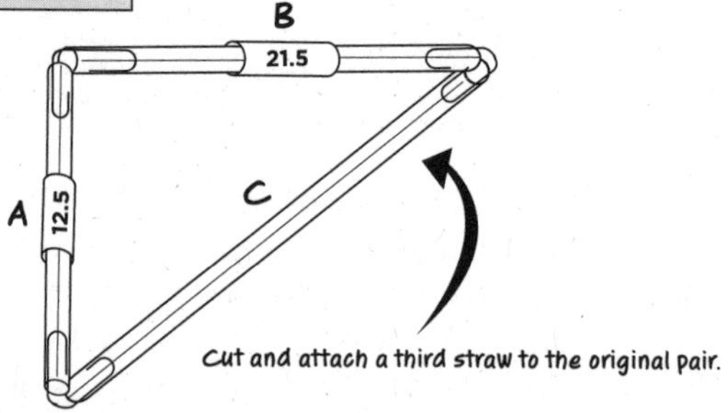

B

21.5

A   12.5

C

Cut and attach a third straw to the original pair.

Next, cut two straws into two different lengths. See **FIGURE 2**.

In the example shown in **FIGURE 3**, the straws are labeled 12.5 centimeters (cm) and 21.5 cm.

Push the right-angle paper clip into the ends of the straws as shown in **FIGURE 4**.

Place another straw near the ends of the other two and cut a length that fits. Attach it to the other two straws with paper clips. See **FIGURE 5**.

FIGURE 6 $A^2 + B^2 = C^2$

FIGURE 7 A^2 = first straw length squared
B^2 = second straw length squared
C^2 = length of longest straw length squared

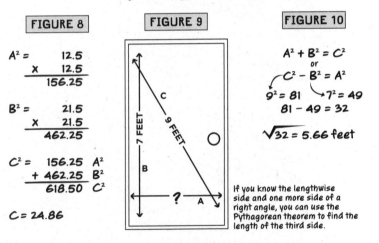

FIGURE 8

$A^2 = $ 12.5
$\times$ 12.5
156.25

$B^2 = $ 21.5
$\times$ 21.5
462.25

$C^2 = $ 156.25 A^2
+ 462.25 B^2
618.50 C^2

$C = 24.86$

FIGURE 9

7 FEET

9 FEET

C

B

? A

FIGURE 10

$A^2 + B^2 = C^2$
or
$C^2 - B^2 = A^2$
$9^2 = 81$ $7^2 = 49$
$81 - 49 = 32$

$\sqrt{32} = 5.66$ feet

If you know the lengthwise side and one more side of a right angle, you can use the Pythagorean theorem to find the length of the third side.

Use the Pythagorean theorem, shown in **FIGURES 6** and **7**, to calculate the length of the longest straw (Side C).

FIGURE 8 shows the calculation needed to find the length of the third, unlabeled straw.

Now you can use the same procedure to determine an unknown length of a rectangle everywhere you go. **FIGURE 9** shows how to perform this on a door. **FIGURE 10** displays the calculation used to find the diagonal measurement.

DETERMINE OBJECT HEIGHTS USING SHADOWS

Sneaky Method 1

Once a day the sun will produce a shadow that equals your height and the height of every other object.

To determine how tall a tree is, for example, simply measure your shadow to determine the time this occurs and then mark off and measure the length of the corresponding shadow of a tall object, like the tree in **FIGURE 1**.

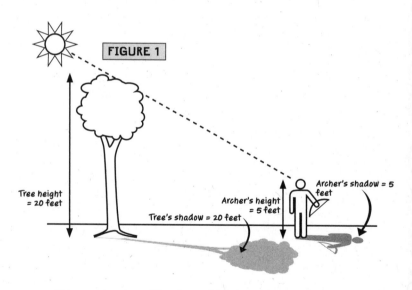

Trigonometry—Shadow / Height Determination

When your shadow length equals your height, measure a tall object's shadow. It will also equal its height.

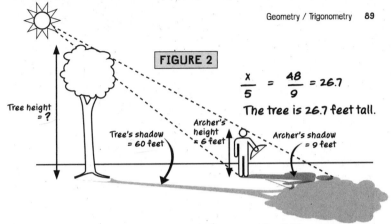

FIGURE 2

$$\frac{x}{5} = \frac{48}{9} = 26.7$$

The tree is 26.7 feet tall.

Tree height = ?

Tree's shadow = 60 feet

Archer's height = 6 feet

Archer's shadow = 9 feet

■ **When your shadow is not equal to your height, measure it and the tall object's shadow. Use the formula below to determine the tall object's height.**

$$\frac{\text{height of archer}}{\text{height of archer's shadow}} = \frac{\text{length of tall object's shadow}}{\text{length of tall object's shadow}}$$

Sneaky Method 2

If you cannot wait to take a shadow measurement when it equals your height, you can measure your shadow and the shadow of the tree and solve for its height with a simple proportion fraction formula.

For instance, if your shadow is 9 feet long and your height is actually 6 feet, measure the shadow of the tree and use the following formula:

YOU: Height = 6 feet   **TREE:** Height = Unknown (X)
          Shadow = 9 feet   Shadow = 60 feet

Proportion Equation $\dfrac{6}{9} = \dfrac{X}{60}$ *(cross-multiply as shown below)*

6 x 60 = 360   9 times X = $9X$ 360 = $9X$
360 / 9 = 40 $X = 40$ The tree's height is 40 feet.

Project—
SNEAKY HYPSOMETER

The Pythagorean theorem allows you to find the height of objects when you know the distance from their base and at least one side and one angle of a triangle the position forms. Astronomers have been able to use trigonometry to find the distances of stars by using the triangle the earth forms when it is in different positions during its rotation around the sun.

You can make a device called a hypsometer to find the heights of tall objects using everyday things.

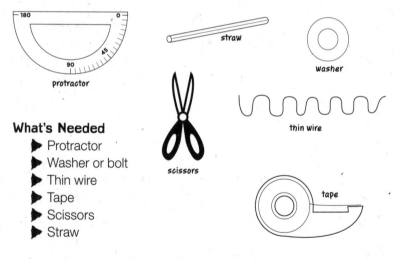

protractor

straw

washer

thin wire

scissors

tape

What's Needed
▶ Protractor
▶ Washer or bolt
▶ Thin wire
▶ Tape
▶ Scissors
▶ Straw

What to Do
Wrap a 1-foot length of wire through the middle of a washer and tape it to the middle of the straight side of the protractor. See **FIGURE 1**.

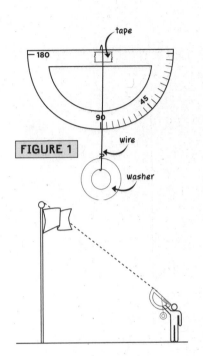

FIGURE 1

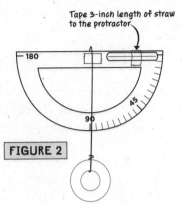

FIGURE 2

Next, cut a 3-inch length of the straw and tape it along the right side of the protractor's straight side. See **FIGURE 2**.

The washer should be able to hang and move freely.

Using the Sneaky Hypsometer

When you know the length of one side and an angle of a right-angle triangle, you can discover the length of the missing side. **FIGURE 1** on the next page shows the 60-degree angle is the tangent angle. According to trigonometry charts, or using a scientific calculator, the tangent number is 1.7321. The boy's viewing height is 5 feet, and the distance to the flagpole, which represents another side of the triangle is 10 feet. Using these figures allows you to calculate the height of the flagpole.

Note: Every angle has its own unique tangent number. A chart is included in the Reference section (see page 177) of the book.

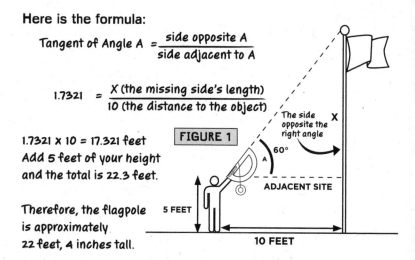

Here is the formula:

$$\text{Tangent of Angle A} = \frac{\text{side opposite A}}{\text{side adjacent to A}}$$

$$1.7321 = \frac{X \text{ (the missing side's length)}}{10 \text{ (the distance to the object)}}$$

1.7321 x 10 = 17.321 feet
Add 5 feet of your height
and the total is 22.3 feet.

FIGURE 1

The side opposite the right angle   X

60°
A

ADJACENT SITE

5 FEET

Therefore, the flagpole
is approximately
22 feet, 4 inches tall.

10 FEET

As an example, in **FIGURE 1** the boy is shown 10 feet away from a flagpole. His height while looking through the Sneaky Hypsometer's straw sight was 5 feet.

Go outside and face the direction of a nearby tall tree or building. Hold the Sneaky Hypsometer at eye level and, using your height as a guide, estimate how high the hypsometer is from the ground. (Eyes are typically 5–8 inches below the top of your head, so keep this in mind when guessing the height of the hypsometer from the ground.)

Next, go up to the tall object and move away from it while keeping track of the distance you retreat. You should be at least 10 feet away to obtain a good height reading.

Look up at the top of the tall object through the hypsometer's straw, and note the angle on the protractor where it passes the scale to calculate the object's height using the formula above.

■ Trigonometry Fundamentals

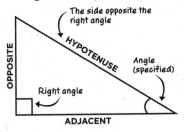
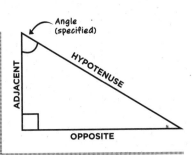

■ Trigonometry Formulas to Determine Unknown Sides

Tangent angle $= \dfrac{opposite}{adjacent}$

(When the *opposite* and *adjacent* lengths are known.)

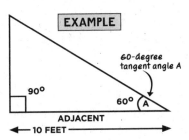

EXAMPLE

60-degree tangent angle A

90°   60°   A

ADJACENT

← 10 FEET →

Cosine angle $= \dfrac{adjacent}{hypotenuse}$

(When the adjacent and hypotenuse lengths are known.)

Sine angle $= \dfrac{opposite}{hypotenuse}$

(When the opposite and hypotenuse are known.)

To remember the proper formula, use the

S O H C A H T O A

mnemonic:
Sine Opposite Hypotenuse Cosine Adjacent
Hypotenuse Tangent Opposite Adjacent

NOTES

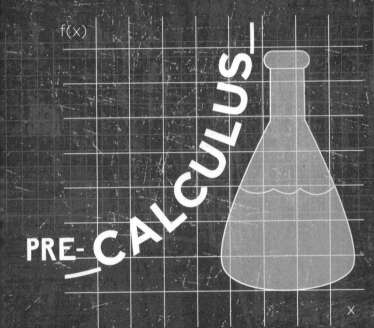

Part 4:

f(x)

PRE-CALCULUS

x

MATH SYMBOLS
DELTA

The delta symbol denotes the change in value.

DRIVE IT!

The vehicle started at location X and is now located at position delta X.

X   △X

MATH SYMBOLS
SUMMATION

The summation symbol, or sum for short (also known as the Greek symbol sigma), represents the operation of adding a sequence of numbers.

DRIVE IT!

Notice the 1 below and the 10 above the sum symbol. With the variable X to the right, this means to add the numbers 1 through 10 to the value of X and sum their total.

$\sum_{1}^{10} X$ If $X = 1$, then the sum operation is as follows:
$1+2+3+4+5+6+7+8+9+10 = 55$

$\sum_{1}^{5} X+1$ In the next example, 1 is below and 5 is above the Sum symbol. To the right of it is $X + 1$.

$\sum_{1}^{5} X+1$ Starting at $X = 1$, X is added to the number 1 each time, and continuing to $X = 5$:
$(1+1) + (2+1) + (3+1) + (4+1) + (5+1) = 20$

MATH SYMBOLS
INTEGRAL

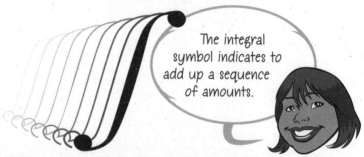

The integral symbol indicates to add up a sequence of amounts.

■ The sum symbol means add up a specific number of finite amounts, whereas the integral sign allows infinitely small amounts to be added together.

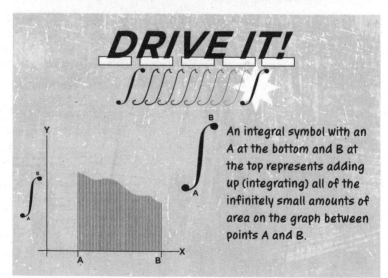

DRIVE IT!

An integral symbol with an A at the bottom and B at the top represents adding up (integrating) all of the infinitely small amounts of area on the graph between points A and B.

CALCULUS

■ **Calculus is the "mathematics of change."**
It consists of two branches, differential and integral calculus.
The following section covers the fundamentals of differential
calculus. With differential calculus you can determine the rate
of change for problems that are continually evolving, such as
velocity, growth rate, or volume level.

■ **Determining the height of water filling a cylindrical jar at
a steady rate is easy. You can predict the water level at
any point in time using an algebra slope formula:**

$$slope = \frac{rise}{run}$$

■ **Finding the *changing rate* of
the water level while filling
a bell-shaped jar
requires calculus.**

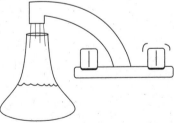

Project—
CALCULUS VS. ALGEBRA RATE OF CHANGE DEMO

The following project provides an easy way to demonstrate why calculus is needed to calculate changing rates.

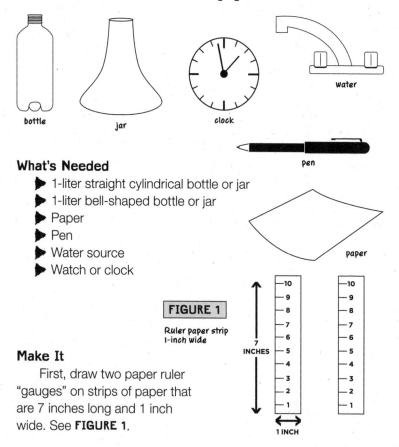

bottle   jar   clock   water   pen   paper

What's Needed
- 1-liter straight cylindrical bottle or jar
- 1-liter bell-shaped bottle or jar
- Paper
- Pen
- Water source
- Watch or clock

FIGURE 1

Ruler paper strip
1-inch wide

Make It
First, draw two paper ruler "gauges" on strips of paper that are 7 inches long and 1 inch wide. See **FIGURE 1**.

Affix the paper strips to the two bottles at the same position in height, and apply tape to hold them in place, as shown in **FIGURE 2**.

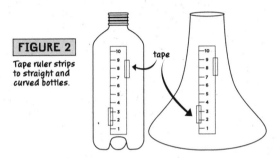

FIGURE 2

Tape ruler strips to straight and curved bottles.

tape

Next, draw a Straight vs. Curved Bottle Fill Chart, as shown in **FIGURE 3**.

FIGURE 3	Straight Bottle Time	Curved Bottle Time
Level 1		
Level 2		
Level 3		
Level 4		
Level 5		
Level 6		
Level 7		
Level 8		
Level 9		
Level 10		

NOTE: You may want to perform this procedure with a friend. One of you can fill the bottle and tell the other when certain levels are reached. The other person can keep track of time and mark the fill chart.

Next, turn the water faucet to a level that produces a slow and steady flow. Mark the faucet's position so you can duplicate the water flow rate later on. Set your stopwatch to zero or wait until the beginning of the next minute, and place the straight bottle under the faucet, as shown in **FIGURE 4**.

Mark down the time that the water level reaches each line on the paper strip until the bottle is full. You can also create a graph to illustrate the results, as shown in **FIGURE 5**.

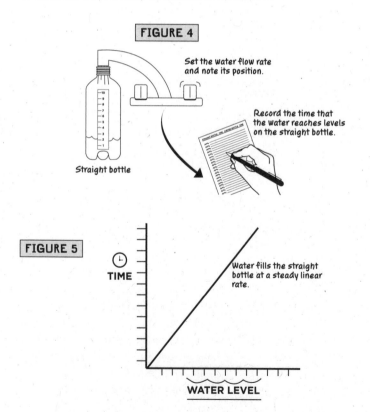

FIGURE 4

Set the water flow rate and note its position.

Record the time that the water reaches levels on the straight bottle.

Straight bottle

FIGURE 5

TIME

Water fills the straight bottle at a steady linear rate.

WATER LEVEL

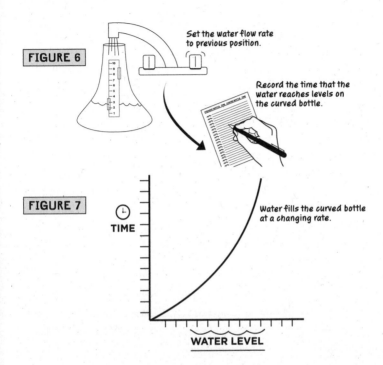

FIGURE 6

Set the water flow rate to previous position.

Record the time that the water reaches levels on the curved bottle.

FIGURE 7

TIME

Water fills the curved bottle at a changing rate.

WATER LEVEL

Then, reset your stopwatch to zero or wait until the beginning of the next minute on your watch, and place the bell-shaped bottle under the faucet, as shown in **FIGURE 6**. Mark down the time that the water level reaches each line on the paper strip until the water fills the bottle.

Even though the bottles have the same volume capacity and the water was pouring at the same rate, the water level was predictable for the straight bottle but not so much for the curved one. This makes sense because the bell-shaped bottle has more volume near the bottom compared to the narrow top section. See **FIGURE 7**.

You can also graph your results for a visual presentation.

CALCULUS AND DETERMINING THE RATE OF CHANGE

- A changing rate appears as a curved line on a graph. By finding the slope of a TANGENT LINE to the curve, you can determine the INSTANTANEOUS RATE of filling (or whatever you are measuring) at that point.

- When you zoom in close enough, a curved line appears straight. A second point, extremely close to, but not touching the first point, allows you to calculate the slope of the curve at a single point. A LIMIT prevents the space between the two points from reaching zero (because you cannot divide by zero).

- A LIMIT is like a doorstop that keeps the door from touching the wall. The symbol looks like this:

$$\lim_{h \to 0}$$

It states that the variable h can get infinitely close to but never reach zero.

CALCULUS LIMITS

Car gets very close to wall (but doesn't touch it).

Boat gets as close as possible to dock (without touching it).

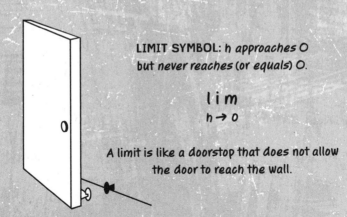

LIMIT SYMBOL: h approaches 0 but *never* reaches (or equals) 0.

$$\lim_{h \to 0}$$

A limit is like a doorstop that does not allow the door to reach the wall.

◼ Derivative of a Curve

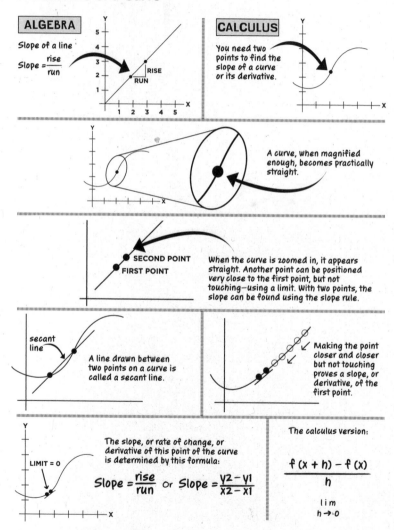

ALGEBRA

Slope of a line

$$Slope = \frac{rise}{run}$$

CALCULUS

You need two points to find the slope of a curve or its derivative.

A curve, when magnified enough, becomes practically straight.

SECOND POINT
FIRST POINT

When the curve is zoomed in, it appears straight. Another point can be positioned very close to the first point, but not touching—using a limit. With two points, the slope can be found using the slope rule.

secant line

A line drawn between two points on a curve is called a secant line.

Making the point closer and closer but not touching proves a slope, or derivative, of the first point.

LIMIT = 0

The slope, or rate of change, or derivative of this point of the curve is determined by this formula:

$$Slope = \frac{rise}{run} \quad or \quad Slope = \frac{y2 - y1}{x2 - x1}$$

The calculus version:

$$\frac{f(x + h) - f(x)}{h}$$

$$\lim_{h \to 0}$$

Project—
CALCULUS DIFFERENTIALS
AND SLOPE DEMONSTRATION

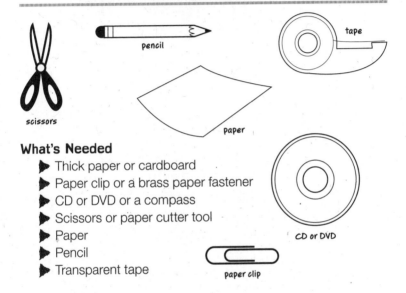

What's Needed
- Thick paper or cardboard
- Paper clip or a brass paper fastener
- CD or DVD or a compass
- Scissors or paper cutter tool
- Paper
- Pencil
- Transparent tape

You can use the illustrations on the next pages as a guide or photocopy the pages, paste them on the cardboard, and cut out the shapes shown in **FIGURES 1** through **6**.

Trace a disc the size of a CD or DVD, or use a compass to create a circle with a diameter of $4\frac{3}{4}$ inches, as shown in **FIGURE 1**.

Next, cut out the rectangular cover shown in **FIGURE 2**, including the "window" holes, or you can rub a pen firmly along the dotted lines to cut away the window section.

Fold the cover in half and puncture a hole in the front. See **FIGURE 3**. Also puncture a hole in the center of the disc, as shown in **FIGURE 4**. Place the disc in the cover and carefully push the

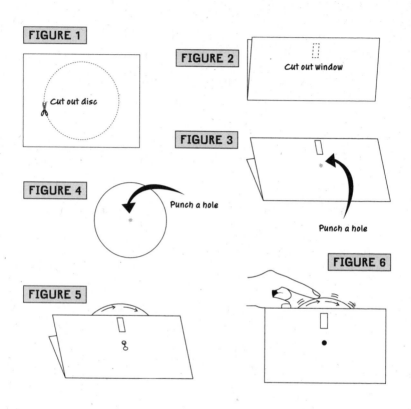

FIGURE 1

Cut out disc

FIGURE 2

Cut out window

FIGURE 3

Punch a hole

FIGURE 4

Punch a hole

FIGURE 5

FIGURE 6

brass paper fastener or paper clip from the outside of the cover and through the disc. Place the disc in the cover and carefully push the brass paper fastener or paper clip from the outside of the cover and through the disc, as shown in **FIGURE 5**.

Review the calculus primer information on the front and back of the card and then dial the disc around to see the second point meet the first and feel the limit in action. See **FIGURE 6**.

Enlarge on a copier 125%.

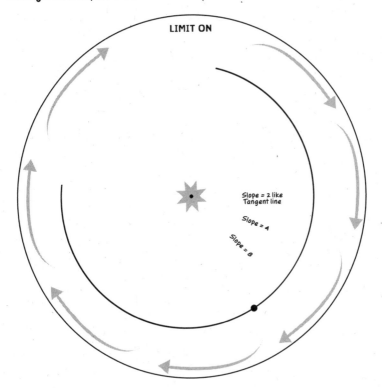

LIMIT ON

Slope = 2 like
Tangent line

Slope = 4

Slope = 8

NOTE: After reviewing the principles and techniques in *Sneaky Math*, like variables, functions, and Cartesian plane graphs, you may want to see what techniques are presented in a calculus text or class. See "Going Further—Calculus" on page 127 for additional material.

CALCULUS
IS THE MATHEMATICS OF CHANGE.

Turn the dial to view the three points on the curve approach each other and see the LIMIT in action.

With calculus you can determine the rate of change of problems that are continually evolving, such as velocity, growth rate, or volume level.

f(x)

WATER LEVEL

cut along dotted line

cut along dotted line

tangent line

TIME

x

Function $f(x) = x^2$

Notice the fill rate of the bell-shaped bottle and the function curve are similar—slow at the bottom and faster near the top.

CALCULUS

SNEAKY MATH:
A Graphic Primer with Projects

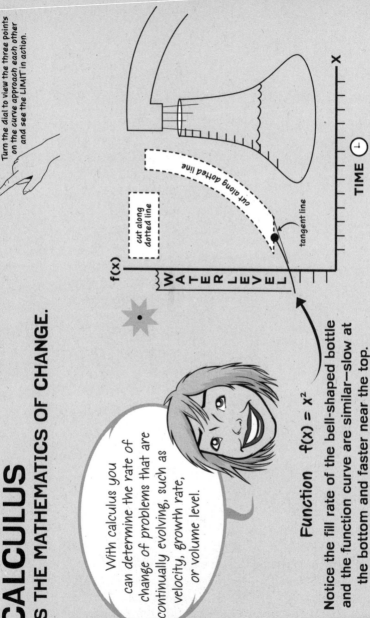
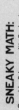

SNEAKY CALCULUS

Determining the steady rate of water filling a straight jar is easy. You can predict its water level at any point in time using an algebra slope formula.

Finding the changing rate of water filling a bell-shaped jar requires calculus.

When you zoom in close, a curved line appears straight. A second point, extremely close to, but not touching, the first point allows you to calculate the slope of the curve at a single point. A limit prevents the two points from reaching zero (because you cannot divide by zero).

CURVED LINE APPEARS STRAIGHT.

2nd POINT

A changing rate appears as a curved line. Finding the slope of a tangent line on the curve allows you to determine the instantaneous rate at that point.

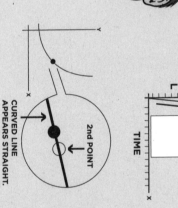

DOOR STOP/LIMIT

A limit is like a doorstop that keeps the door from touching the wall.

The symbol for a limit looks like this:

$$\lim_{h \to 0}$$

It states that the variable h can approach but never reach zero.

TANGENT LINE

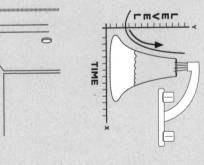

TIME

TIME

NOTES

Part 5:

GOING FURTHER
WITH SNEAKY MATH

The following material will help parents
and teachers and anyone who wants
to go beyond the basics and advance
in math. You'll learn a little more about
algebra and calculus, how to get
started with scientific calculators and
spreadsheets, and how to make math
craft projects with everyday things.

SIGMA
SUMMATION
EXAMPLE

SCIENTIFIC CALCULATORS

Since scientific calculators are inexpensive—free versions can be accessed on the Web—you should learn how to use them as soon as possible. They allow you to take your fundamental math skills and perform time-saving, advanced operations and, in some instances, see a graphical display of your work.

They are not scary at all. This section will introduce you to the basics, test equations, and functions shown earlier in this book and show you how to go much further.

Take math to the next level with a scientific calculator.

FIGURE 1

Simplified view (without decimal buttons).

Things to look for in your scientific calculator:

- With all the following suggestions and tips, it is recommended to get familiar with the exact keys and function options on your own calculator for particular operations.

- Special function key access: Most keys can perform multiple functions. You access these by pressing another key first, usually called 2nd or Shift or Alpha. See **FIGURE 1**.

■ Understand the difference in the (–) and the – keys. The first is for negative numbers. The second is for performing subtraction.

■ Learn how to store variables for use in formulas.

■ Learn how to use tables of numbers for entering function equivalents like f(x) to x.

EXAMPLES:

■ Locate fundamental keys that you will frequently access, such as the Powers key (for exponents), Square Root, and Log. See **FIGURE 2**.

FIGURE 2

Use the Power key x^2 to multiply a number by itself:

$3^2 = 3 \times 3 = 9$

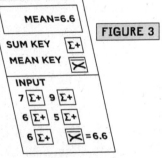

FIGURE 3

Calculate the mean, or average, of a list of numbers.

■ Find the statistics keys, such as Mean, and learn how to use it. In some instances you must input your numbers and select the Sum key. On other models you enter your data values in a table and then select Mean. See **FIGURE 3**.

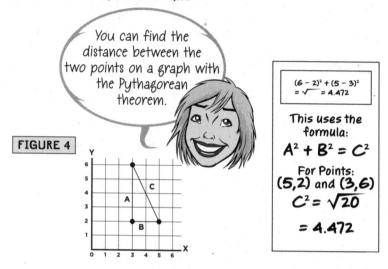

You can find the distance between the two points on a graph with the Pythagorean theorem.

$(6 - 2)^2 + (5 - 3)^2$
$= \sqrt{} = 4.472$

This uses the formula:

$A^2 + B^2 = C^2$

For Points:
(5,2) and (3,6)
$C^2 = \sqrt{20}$
$= 4.472$

FIGURE 4

- **FIGURE 4** includes a useful method to use your calculator to find the distance between two points on a graph using the Pythagorean theorem.

- While the Mean, or Average, function provides information about a data set, you may want to also check the standard deviation of the group. This supplies the "spread" of a group of numbers from the average, which provides insights into group quality and performance. See **FIGURE 5**.

FIGURE 5

Calculate the standard deviation of the spread of a group of numbers.

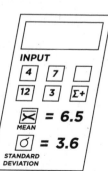

■ You can calculate the distance and time for an object to fall, barring wind resistance, with the free-fall formula. To find the height an object was released when you know the time of the fall, use this formula:

$$\frac{1}{2} AT^2 = D$$

A = acceleration rate (32 feet per second)
T = time (2 seconds in this example)
D = distance (height the object was dropped)

With this information, calculate the free-fall drop as follows:

$$\frac{1}{2} A \text{ times } T^2 = 16 T^2 = 16 \text{ times } 2^2 = 64$$

The object, which took two seconds to hit the ground, was dropped from a height of 64 feet (19.51 meters). See **FIGURE 6**.

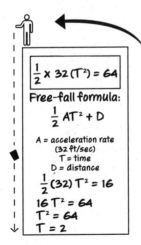

FIGURE 6

$$\frac{1}{2} \times 32 (T^2) = 64$$

Free-fall formula:
$$\frac{1}{2} AT^2 + D$$

A = acceleration rate
(32 ft/sec)
T = time
D = distance
$$\frac{1}{2} (32) T^2 = 16$$
$$16 T^2 = 64$$
$$T^2 = 64$$
$$T = 2$$

How high was the object when it was released? In two seconds the item fell approximately 64 feet or 19.51 meters.

When you know the height, calculate the time to fall.

Height = 30 feet

$$T = \sqrt{\frac{2D}{A}}$$

T = time
D = distance
A = acceleration rate

$$\sqrt{\frac{2 \times 30 \ (ft)}{32 \ (ft \ per \ second)}}$$

$$= \sqrt{\frac{60}{32}}$$

$$= \sqrt{1.88} = 1.37$$

The item takes 1.37 seconds to hit the ground.

FIGURE 7

←— GROUND ——

■ Conversely, if you know the height of a falling object, you can calculate the time of the fall.

FORMULA:

$$T = square \ root \ of \ (2 \times \frac{D}{A})$$

Time = time to fall
D = distance (30 feet)
A = acceleration rate (32 feet per second)

FIGURE 7 shows that the object dropped from 30 feet takes 1.37 seconds to hit the ground.

■ Your scientific calculator really comes in handy when figuring out complex formulas like the wind-chill index.

Enter the formula for the wind-chill index below.

FORMULA:

$$35.74 + (0.6251)\,T - (35.75)\,V^{0.16} + (0.4275)\,T\,V0.16$$

T = temperature in Fahrenheit
V = wind velocity in miles per hour

FIGURE 8 shows how to calculate the wind chill for a temperature of 40 degrees with a wind speed of 10 mph:

35.74 + (0.6251) (40 degrees) − (35.75) (10 mph) 0.16 + (0.4275) (40 degrees) (10 mph) 0.16 = 35.74 + 24.86 (35.75) (1.44) + (0.4275) (40) (1.44) = 60.6 − 51.48 + 24.62 wind chill of 33.74 degrees

FIGURE 8   T = 40°   V = 10 mph

The wind-chill index = 33.64° F

- Algebra functions can be generated for you if your calculator includes a Table function.

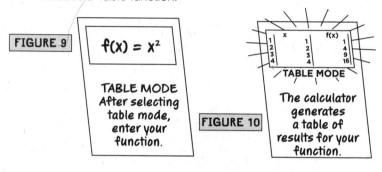

FIGURE 9

$f(x) = x^2$

TABLE MODE
After selecting
table mode,
enter your
function.

FIGURE 10

x | f(x)
1 | 1
2 | 2 | 4
3 | 3 | 9
4 | 4 | 16

TABLE MODE

The calculator
generates
a table of
results for your
function.

FIGURE 9 shows a calculator using the Table function. The beginning of the function is displayed:

$f(x) =$

You then enter your function of X.
In this example enter X^2 (X squared).

The calculator will display 2 columns.
The left one shows the numbers 1 through 4.

The right column displays the result of the
function of X, which equals X squared.

Continue experimenting with various formulas and functions with your scientific calculator so when they are required for class you will be skilled in their use.

Before we present the methods to solve differential and integral calculus equations, you should review the following two sections. You will learn how to expand and simplify algebraic terms and how to use the power rule in calculus to save time.

ALGEBRA— Expanding and Simplifying Terms

An algebraic technique called simplifying and expanding terms is performed in many math equations. It is like reducing fractions— the terms have the same values but it's easier to complete calculations with other terms.

EXAMPLE 1:

FIGURE 1 $(x + 3)^2$ Expanding is the process of removing the parentheses (so you can complete a calculation).

Expanding is the process of removing the parenthesis from a term like $(x + 3)^2$ shown in **FIGURE 1**.

When expanded, $(x + 3)^2 = (x + 3)(x + 3)$, as shown in **FIGURE 2**.

The *sum* of a variable and number, squared.

FIGURE 2

$$(x + 3)^2 = (x + 3)(x + 3)$$

$$(x + 3) \text{ times } (x + 3)$$

FIGURE 3 To expand further . . . multiply the items in the left parentheses times the items in the right parentheses.

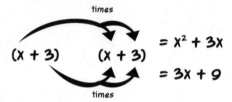

$(x + 3) \quad (x + 3) \quad = x^2 + 3x$
$= 3x + 9$

FIGURE 4 Let's multiply one step at a time:

First
$(x + 3) \xrightarrow{times} (x \quad = x^2$

Outside
$(x + 3) \xrightarrow{times} 3) = 3x$

Inside
$(x + 3) \xrightarrow{times} (x \quad = 3x \quad = x^2 + 3x + 3x + 9$

Last
$(x + 3) \xrightarrow{times} 3) = 9 \quad = x^2 + 6x + 9$

To expand the term further, you can multiply the first term in the left parenthesis by the first one in the second parenthesis. See **FIGURE 3**. **FIGURE 4** shows the First, Outside, Inside, and Last—or FOIL—technique to multiply terms.

The original term $(x + 3)^2$ expands to $x^2 + 6x + 9$ as shown in **FIGURE 5**.

FIGURE 5

Expanding and simplifying terms allows you to work with other terms—similar to finding a common denominator with mixed fractions.

$$So, (x + 3)^2$$

EXPANDS to:
$$(x + 3) \text{ times } (x + 3)$$

Which EXPANDS to
$$x^2 + 6x + 9$$

EXAMPLE 2:

In some situations you may need to simplify a long term into a shorter one to complete a calculation. For instance, the term $2xh + h^2$ simplifies to $h(2x + h)$.

To prove it, multiply $h \times 2x$ to produce $2hx$. Then multiply $h \times h$ to get h^2 so their sum is equal to $2hx + h^2$. See **FIGURE 6**.

FIGURE 6

Here's how to simplify a term with two variables and one number shared.

$$2xh + h^2 = h(2x + h)$$

times
$$h (2x + h) = 2hx$$

times
$$h (2x + h) = h^2$$

$$= 2hx + h^2$$

Simplifying is like reducing a fraction.
It has the same value, but it becomes easier to work with.

CALCULUS— THE POWER RULE

There are rules in calculus that can save you time. One of them is called the power rule.

- You'll learn how to use it when calculating differential or integral equations.

THE POWER RULE FOR DIFFERENTIALS

The power rule applies to functions with a power notation, like the variable x^2 (x to the power of 2). To save time finding the derivative of this function, simply place the exponent 2 in front of the variable and reduce the exponent by 1.

As shown in **FIGURE 1**, x^2 becomes $2x^1$ or just $2x$.

FIGURE 1

EXAMPLE 1. To find the derivative of function $f(x) = x^2$, place the exponent 2 in front of x and reduce the exponent (2) by 1. The derivative is 2x.

$$2x^{2-1=1} = 2x$$

EXAMPLE 2. $f(x) = x^3$ becomes $3x^2$

$$3x^{3-1=2} = 3x^2$$

THE POWER RULE FOR INTEGRALS

Here is the power rule for evaluating anti-derivatives.

■ When an integral function includes power term, perform the opposite of the technique shown for differentials.

EXAMPLE 1. x^2 becomes $\frac{x^3}{3}$

How? First, increase the exponent value by 1 and divide by that number.

$$x^2 \text{ becomes } x^{2+1} = \frac{x^3}{3}$$

EXAMPLE 2. x^3 becomes $x^{3+1} = \frac{x^4}{4}$

EXAMPLE 3. x^4 becomes $x^{4+1} = \frac{x^5}{5}$

CALCULUS–
Formula for Determining
the Rate of Change

This section will show in detail how a function's rate of change is determined using calculus. To make it easy to understand, a linear function is displayed alongside the constantly changing function for comparison.

FIGURE 1 shows the bell-shaped bottle next to the changing function curve representing the function $f(x) = x^2$.

In the illustrations that follow, the graph's horizontal line is called the x-axis and the vertical line is called the f(x)-axis (instead of the y-axis). **FIGURE 2** shows two points on a graph that form a linear function (or a straight line). On the right, a changing rate is displayed with the function $f(x) = x^2$.

The two points on the linear function are represented as (2,8) and (4,16) on the graph. These points are also represented by variables (x, f(x)) and (x + h, f(x) + h) respectively. The points on the changing function on the right are represented as (2,4) and (4,16) on the graph. These points are also represented by variables (x, f(x)) and (x + h, f(x) + h) respectively. See **FIGURE 3**.

The linear function's rate is determined by the formula

$$\text{Slope} = \frac{\text{RISE}}{\text{RUN}}$$

This is represented as: $\dfrac{f(x + h) - f(x)}{x + h - x}$

GOING FURTHER—CALCULUS:

Formula for Rate of Change

FIGURE 1

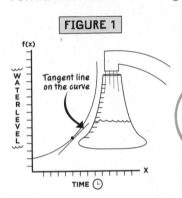

f(x)

WATER LEVEL

Tangent line on the curve

TIME 🕐

x

You saw the basic way an instantaneous rate of change is determined for a constantly changing function.

Now see how to calculate the instantaneous rate at a point compared to a function with a steady, linear rate.

FIGURE 2   In this example, the y-axis is shown as f(x), or function of x

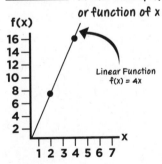

f(x)

16
14
12
10
8
6
4
2

1 2 3 4 5 6 7   x

Linear Function
f(x) = 4x

Examples

x	f(x)
1	4
2	8
3	12
4	16

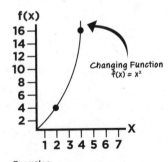

f(x)

16
14
12
10
8
6
4
2

1 2 3 4 5 6 7   x

Changing Function
f(x) = x²

Examples

x	f(x)
1	1
2	4
3	9
4	16

FIGURE 3

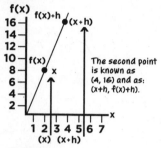

The second point is known as (4, 16) and as: (x+h, f(x)+h).

The first point, (2,8), is also known as point (x, f(x)).

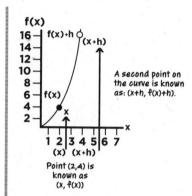

A second point on the curve is known as: (x+h, f(x)+h).

Point (2,4) is known as (x, f(x)).

FIGURE 4

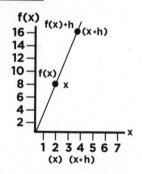

To determine the rate of the linear function f(x)=4x, use the slope formula.

$$\text{Slope} = \frac{rise}{run} \quad \text{or} \quad \frac{f(x + h) - f(x)}{(x + h) - x}$$

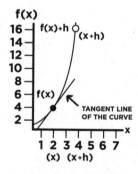

TANGENT LINE OF THE CURVE

To determine the instantaneous rate at point (2,4), use the differential equation to find the slope of the tangent line.

$$\lim_{h \to 0} = \frac{f(x + h) - f(x)}{(x + h) - x}$$

FIGURE 5

Rate Formula: Slope $= \dfrac{rise}{run}$

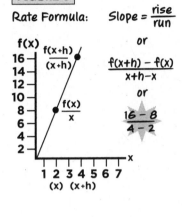

or

$\dfrac{f(x+h) - f(x)}{x+h-x}$

or

$\dfrac{16 - 8}{4 - 2}$

Rate of Change Formula:
This function is equal to $f(x) = x^2$.

$$\lim_{h \to 0} = \frac{f(x + h) - f(x)}{(x + h) - x}$$

Substitute function $f(x) = x^2$.

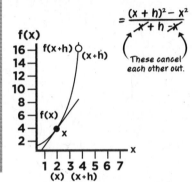

$$= \frac{(x + h)^2 - x^2}{x + h - x}$$

These cancel each other out.

To find the changing function's instantaneous rate of change (which is also the slope of a tangent line to the curve) at point (2,4), the differential equation is used as follows:

$$\lim_{h \to 0} = \frac{f(x + h) - f(x)}{(x + h) - x}$$

The linear function's rise/run points are 16 – 8 / 4 – 2. Over on the changing function to the right, we substitute the values of the function $f(x) = x^2$ into the differential equation variables of lim h→0 = f(x+h) – f(x) / (x + h -x).

This equates to: lim h→0 = (x + h) 2^2 – x^2 / (x + h -x) as shown in **FIGURE 5**.

You can cancel the positive x and -x to leave just h as the denominator.

FIGURE 6

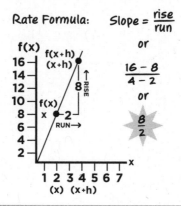

Rate Formula: Slope = $\frac{rise}{run}$

or

$\frac{16 - 8}{4 - 2}$

or

$\frac{8}{2}$

Rate of Change Formula:

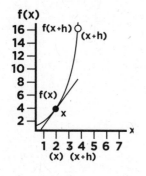

$\lim_{h \to 0} = \frac{(x + h)^2 - x^2}{h}$

$(x+h)^2$ EXPANDS to: $x^2 + 2xh + h^2$

$\lim_{h \to 0} = \frac{x^2 + 2xh + h^2 - x^2}{h}$

These cancel each other out.

The linear equation's fraction term reduces to $\frac{8}{2}$. On the bottom, the numerator $(x + h)^2$ expands to $x^2 + 2xh + h^2$. Now the positive x^2 and the minus x^2 cancel each other, leaving just $2xh + h^2$ as shown in **FIGURE 6**.

FIGURE 7 shows the linear equation's value of $\frac{8}{2}$ reduces to $\frac{4}{1}$ which is equal to 4. The two points on the graph represent a slope and rate of 4.

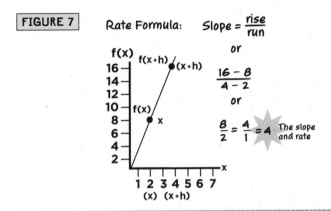

FIGURE 7 — Rate Formula: Slope = $\frac{rise}{run}$

or

$\frac{16-8}{4-2}$

or

$\frac{8}{2} = \frac{4}{1} = 4$ The slope and rate

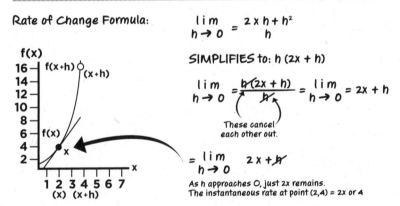

Rate of Change Formula:

$\lim_{h \to 0} = \frac{2xh + h^2}{h}$

SIMPLIFIES to: h (2x + h)

$\lim_{h \to 0} = \frac{h(2x+h)}{h} = \lim_{h \to 0} = 2x + h$

These cancel each other out.

$= \lim_{h \to 0} 2x + h$

As h approaches 0, just 2x remains. The instantaneous rate at point (2,4) = 2x or 4

On the bottom, the term lim h→0 = 2xh + h² can be reduced to just 2x because the h is approaching a near-zero value. At this point, only 2x remains. In conclusion, since the value of x on the graph = 2, the instantaneous rate at point (2,4) = 4.

You can go further and find the instantaneous rates of other points on the graph for the same changing function. For example, calculate the instantaneous rate at point (3,8).

Calculus provides a way to predict the results of model simulations to optimize results.

- You may wonder why so many steps are necessary to calculate a rate of change at a single point on a curve. With a linear (straight-line) function, you can use **any two** coordinate points on the line to find its slope (or rate) because it is **unchanging**.

- However, a function with a changing rate, like $f(x) = x^2$, has no straight line between coordinate points. Calculating a rate at just one point results in zero (because there is no motion or distance traveled).

- A limit allows you to find the instantaneous rate by using a second point on the curve that is an infinitely small distance away from the first point.

- Math functions can be used to develop models of change in position, growth, or decay. Examples include position of planets, compound interest gains, bacteria growth, drug absorption rate, population growth, and radioactive decay.

- Calculating the rate of change and calculus integral problems is much easier with a scientific calculator. The next section shows you how.

CALCULUS– Calculating Differentials with a Scientific Calculator

It is easy to perform calculus differential equations with a properly equipped scientific calculator. The following example will show you how to find the derivative, or slope/rate for function $f(x) = x^2$ at point $x = 2$.

(Always refer to your calculator's user guide for specific key and function selection information.)

First select the Derivative function on the calculator, which appears as a d/dx notation, as shown in **FIGURE 1**.

EXAMPLE 1. Find the differential (rate or slope) for function $(f(x) = x^2)$ at point $x = 2$.

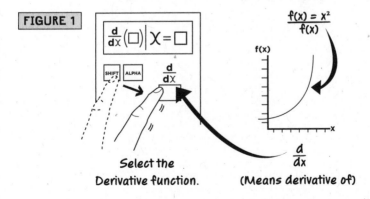

FIGURE 1

Select the Derivative function.

(Means derivative of)

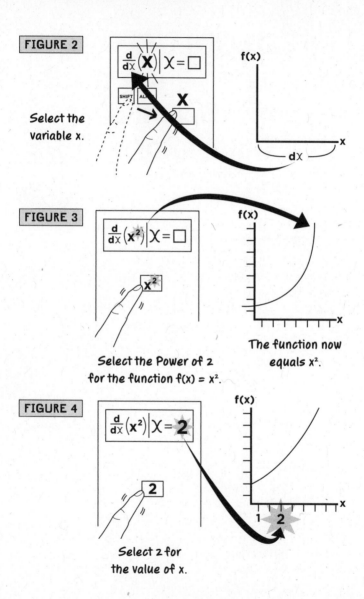

FIGURE 2

$$\frac{d}{dx}(\mathbf{X})\Big|X=\square$$

SHIFT | ALT

Select the variable x.

f(x)

dx

FIGURE 3

$$\frac{d}{dx}(\mathbf{x}^2)\Big|X=\square$$

x^2

Select the Power of 2 for the function f(x) = x².

f(x)

The function now equals x².

FIGURE 4

$$\frac{d}{dx}(\mathbf{x}^2)\Big|X=\mathbf{2}$$

2

f(x)

1 2

Select 2 for the value of x.

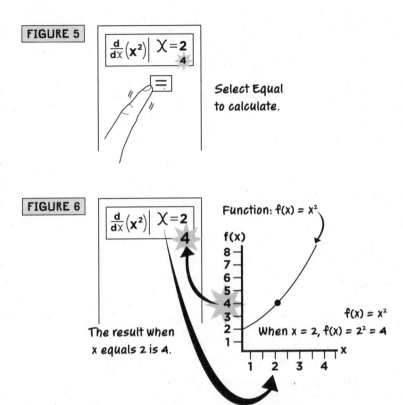

FIGURE 5

$\frac{d}{dx}(x^2)\Big| \; x = 2$

Select Equal to calculate.

FIGURE 6

$\frac{d}{dx}(x^2)\Big| \; x = 2$

4

The result when x equals 2 is 4.

Function: f(x) = x²

f(x)

8
7
6
5
4
3
2
1

When x = 2, f(x) = 2² = 4

f(x) = x²

x

1 2 3 4

Next, select x as the function variable. See **FIGURE 2**.

Then select the Power of 2 for the variable x as shown in **FIGURE 3**. This sets up the problem function f(x) = x².

Move the cursor over to the x = section and select number 2. (Will vary by calculator.) See **FIGURE 4**.

Last, select the Equal sign (=) to calculate. See **FIGURE 5**.

As shown in **FIGURE 6**, the calculator displays the number 4 as the answer.

INTEGRAL CALCULUS

Finding the area of plane figures with straight lines or circles can be performed with regular geometric formulas. Finding the area of curved figures requires integral calculus.

To find the area under a curve, add (or integrate) the lengths and widths of infinitely thin rectangles to form a sum total.

Here's how it's done:

You can approximate the area of a shape that has a curved side by adding together the area of rectangles inside it. See **FIGURE 1**. But this method is not precise.

Using a limit to ensure that the widths of the rectangles are infinitely thin but never reach zero, you can attain a more precise measurement. See **FIGURE 2**.

When you add, or integrate, the height and width of the rectangles, you will find the exact area under the curve, shown in **FIGURE 3**.

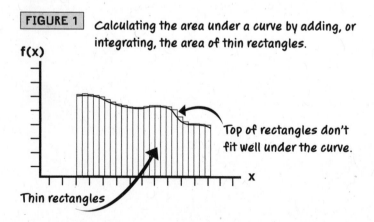

FIGURE 1 Calculating the area under a curve by adding, or integrating, the area of thin rectangles.

Top of rectangles don't fit well under the curve.

Thin rectangles

FIGURE 2 Use a limit to make the width of the rectangles under the curve approach 0.

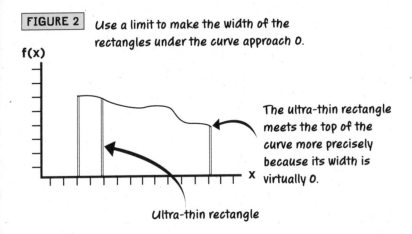

f(x)

The ultra-thin rectangle meets the top of the curve more precisely because its width is virtually 0.

x

Ultra-thin rectangle

FIGURE 3 INTEGRATION

By adding, or integrating, all the ultra-thin rectangles' areas together, you get the area under the curve (area can represent distance, profit, or resource depletion).

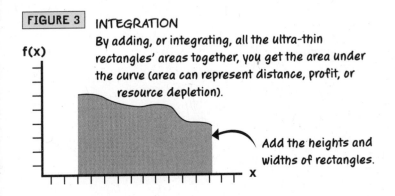

f(x)

Add the heights and widths of rectangles.

x

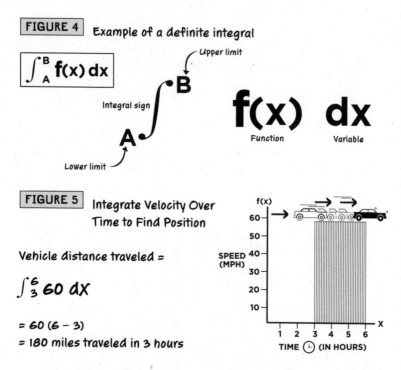

FIGURE 4 Example of a definite integral

$$\int_A^B f(x)\, dx$$

Upper limit

Integral sign

B

A

f(x) dx

Function Variable

Lower limit

FIGURE 5 Integrate Velocity Over Time to Find Position

Vehicle distance traveled =

$$\int_3^6 60\, dx$$

= 60 (6 − 3)
= 180 miles traveled in 3 hours

FIGURE 4 shows an example of a definite integral formula. The integral sign includes an A for the lower limit value and a B for the upper limit value. Next to it, the function is shown with a variable notation following it.

Here's a simple problem you can solve with integration. If your vehicle is moving at a constant speed of 60 miles per hour (mph), how far will it be from its starting point in 3 hours?

In this example, shown in **FIGURE 5**, we will start tracking the car's journey from 3:00 to 6:00. The integral will have a lower limit of 3 (for 3:00) and an upper limit of 6 (6:00). The function, in this instance, is a constant value of 60 (mph).

Integration is the inverse of differentiation. (Some integral functions are called antiderivatives.) See **FIGURE 6**.

FIGURE 6

With integration, you add infinitely small changing amounts to calculate a total change. With differentiation you divide infinitely small changing amounts to calculate a rate of change.

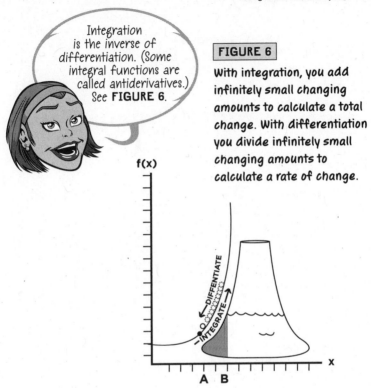

To sum, or integrate the value, and the car's final position, multiply 60 times (6-3), which represents 60 mph times 6 hours minus 3 hours. The vehicle has covered 180 miles from its starting point after traveling at a constant speed for 3 hours.

NOTE: You could have used simple arithmetic to get this result, but this example is intended to show you the integral formula and where to place your data. This will help you with more difficult problems that involve variables with powers (exponents).

CALCULUS– Calculating Integrals with a Scientific Calculator

EXAMPLE 1

You can calculate integral equations with a properly equipped scientific calculator. The following example will show you how to find the area under the curve of point 3 on the y-axis from points 0 and 3 on the x-axis.

(Always refer to your calculator's user guide for specific key and function selection information.)

Select the Integral function as shown in **FIGURE 1**.

EXAMPLE 1. Find the area under the y-axis at 3 and from 0 to 3 on the x-axis.

FIGURE 1

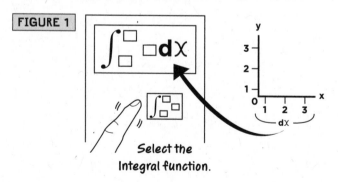

Select the Integral function.

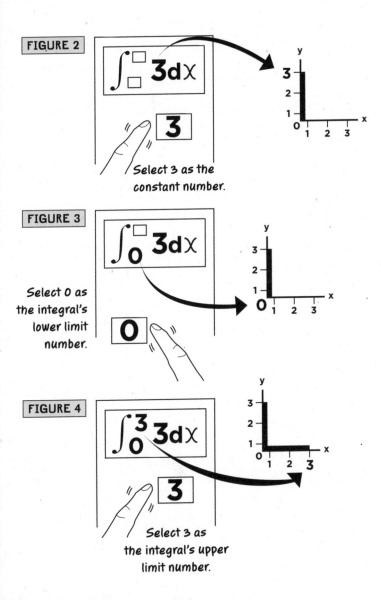

FIGURE 2

$\int_{\square}^{\square} 3dx$

3

Select 3 as the constant number.

FIGURE 3

$\int_{0}^{\square} 3dx$

Select 0 as the integral's lower limit number.

0

FIGURE 4

$\int_{0}^{3} 3dx$

3

Select 3 as the integral's upper limit number.

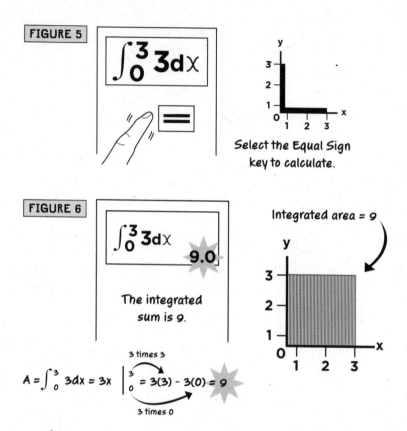

FIGURE 5

$$\int_0^3 3dx$$

Select the Equal Sign key to calculate.

FIGURE 6

$$\int_0^3 3dx$$ 9.0

The integrated sum is 9.

Integrated area = 9

$$A = \int_0^3 3dx = 3x \Big|_0^3 = 3(3) - 3(0) = 9$$

3 times 3

3 times 0

Then input number 3 as the constant number value. See **FIGURE 2**.

Select 0 as the lower limit of the integral, as shown in **FIGURE 3**. Then select 3 for the upper value. See **FIGURE 4**.

Now you're ready to calculate the integral by pressing the Equal Sign (=) key, as shown in **FIGURE 5**.

FIGURE 6 shows the calculated result of 9 as the integrated value. The area formula is also displayed for review.

EXAMPLE 2. Find the area under the curve of function f(x) = x² from 0 to 2.

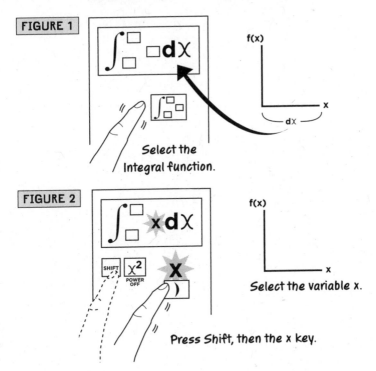

FIGURE 1

Select the Integral function.

FIGURE 2

Select the variable x.

Press Shift, then the x key.

EXAMPLE 2

This problem will demonstrate the real power of integration by calculating a power function. You'll see how to find the area of a curve of function $f(x) = x^2$ from points 0 to 2 on the x-axis.

First select the Integral function. See **FIGURE 1**.

Select the variable x, as shown in **FIGURE 2**.

Next select the Power of 2 button, as shown in **FIGURE 3**.

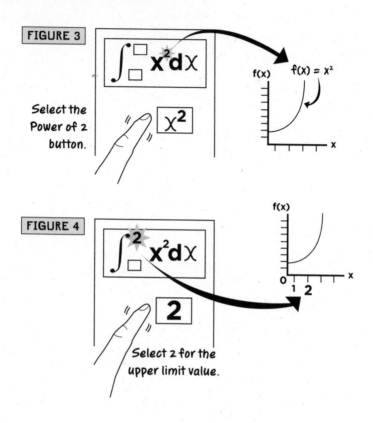

FIGURE 3

Select the Power of 2 button.

$$\int_{\square}^{\square} x^2 dx$$

$f(x)$ $f(x) = x^2$

FIGURE 4

$$\int_{\square}^{2} x^2 dx$$

Select 2 for the upper limit value.

Move the cursor to the upper limit area and input number 2. See **FIGURE 4**.

Move down to the lower limit area and select number 0, as shown in **FIGURE 5**.

Press the Equal Sign key (=) and the calculator sums, or integrates, the area under the curve as $2\frac{2}{3}$. The area formula is also displayed for review.

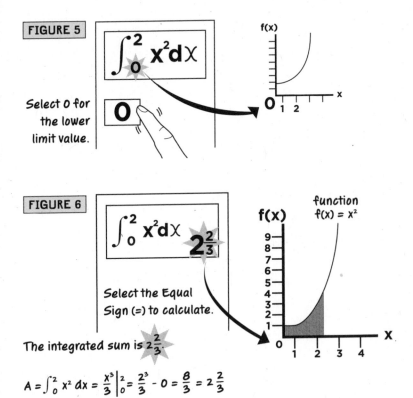

FIGURE 5

Select 0 for the lower limit value.

FIGURE 6

Select the Equal Sign (=) to calculate.

The integrated sum is $2\frac{2}{3}$.

$$A = \int_0^2 x^2\, dx = \frac{x^3}{3}\Big|_0^2 = \frac{2^3}{3} - 0 = \frac{8}{3} = 2\frac{2}{3}$$

The area under the curve of function $f(x) = x^2$ from point 0 to point 2

NOTE: Integration is the inverse of differentiation. (Some integral functions are called antiderivatives.)

With integration, you add infinitely small changing amounts to calculate a total change.

With differentiation, you divide infinitely small changing amounts to calculate rate of change. See **FIGURE 6**.

Bonus Project—
INTEGRAL MATH CARD

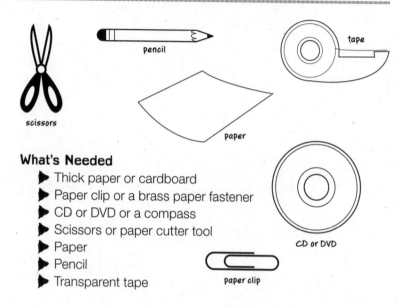

What's Needed
- Thick paper or cardboard
- Paper clip or a brass paper fastener
- CD or DVD or a compass
- Scissors or paper cutter tool
- Paper
- Pencil
- Transparent tape

You can use the illustrations on the next pages as a guide or photocopy the pages, paste them on the cardboard, and cut out the shapes shown in **FIGURES 1** through **6**.

First, trace a disc the size of a CD or DVD, or use a compass to create a circle with a diameter of $4\frac{3}{4}$ inches, as shown in **FIGURE 1**.

Next, cut out the rectangular cover shown in **FIGURE 2** including the "window" holes for the upper limit, area solution, and the triangular "area under the curve." Or, you can rub a pen firmly along the dotted lines to cut away the window section.

Fold the cover in half and puncture a hole in the front, as shown in **FIGURE 3**. Also puncture a hole in the center of the disc,

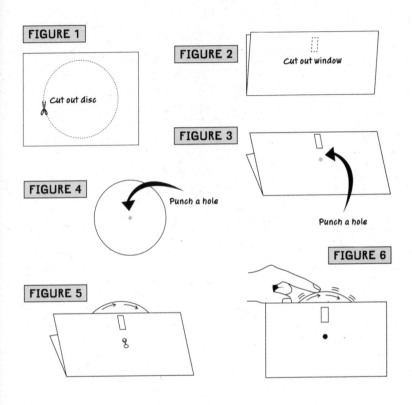

FIGURE 1

Cut out disc

FIGURE 2

Cut out window

FIGURE 3

FIGURE 4

Punch a hole

Punch a hole

FIGURE 6

FIGURE 5

as shown in **FIGURE 4**. Place the disc in the cover and carefully push the brass paper fastener or paper clip from the outside of the cover and through the disc, as shown in **FIGURE 5**.

Dial the disc counter-clockwise, select an integral upper limit number, and see the equation that will match the area shown under the curve. See **FIGURE 6**.

$$AREA = \int x^2 \, dx$$

upper limit

lower limit

function

variable

AREA =

cut along dotted lines

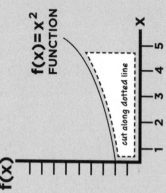

$f(x) = x^2$
FUNCTION

f(x)

x

cut along dotted line

1 2 3 4 5

Turn the dial to check the area under the curve.

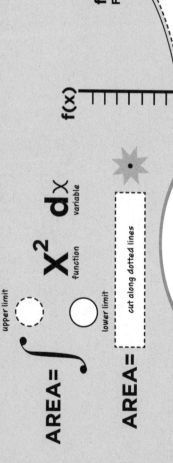

CALCULUS INTEGRALS

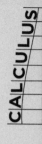

CALCULUS

SNEAKY MATH:
A Graphic Primer with Projects

CALCULUS INTEGRALS

By adding, or integrating, the area of ultra-thin rectangles, you can calculate the area of a function's curve. "Area" can represent physical area, profit gain, decay loss, etc.

Definite Integral Formula

$$A = \int_A^B f \, dx$$

- UPPER LIMIT
- LOWER LIMIT
- FUNCTION
- VARIABLE
- Integral

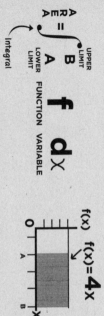

$$f(x) = 4x$$

EXAMPLE 1

"What distance does a vehicle travel that is going 60 mph from 3:00 to 6:00?"

Equation:

$$\int_3^6 60 \, dx$$

- $\int_3^6$ → 3:00 to 6:00
- 60 → 60 miles per hour
- dx → variable X

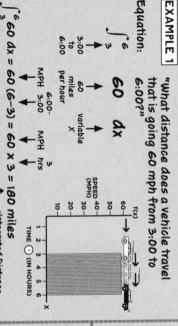

SPEED (MPH)
TIME (IN HOURS)

$$\int_3^6 60 \, dx = 60 \,(6-3) = 60 \times 3 = 180 \text{ miles}$$

- 6:00 → 6:00 → MPH → 3 hrs
- 3:00 → 3:00- → MPH

Integrated Distance

EXAMPLE 2

"Integrate the area under the curve of function f(x) becomes x² from points 0 to 2."

$$f(x) = x^2$$

FUNCTION

$$A = \int_0^2 x^2 \, dx$$

Answer: $\int_0^2 x^2 \, dx = \dfrac{x^3}{3} \Big|_0^2 = \dfrac{2^3}{3} - 0 = \dfrac{8}{3} = 2.7$

The area under the curve from 0 to 2 = 2.7

WAIT! How did x^2 become $\dfrac{x^3}{3}$? See below.

The Power Rule for Integration

To calculate integral functions with power terms, like X to the power of 2 (x^2), you must increase the power by 1 and divide by that new total.

EXAMPLE 1: x^2 becomes x^{2+1} or $\dfrac{x^3}{3}$

EXAMPLE 2: x^3 becomes x^{3+1} or $\dfrac{x^4}{4}$

Enlarge on a copier 125%.

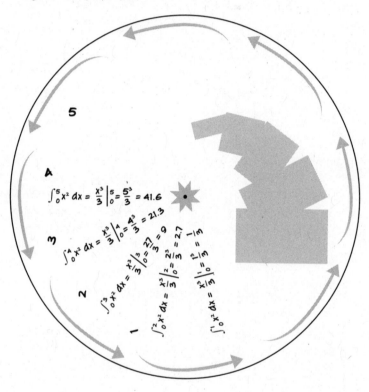

5

4

$\int_0^5 x^2\, dx = \frac{x^3}{3}\Big|_0^5 = \frac{5^3}{3} = 41.6$

3

$\int_0^4 x^2\, dx = \frac{x^3}{3}\Big|_0^4 = \frac{4^3}{3} = 21.3$

2

$\int_0^3 x^2\, dx = \frac{x^3}{3}\Big|_0^3 = \frac{27}{3} = 9$

1

$\int_0^2 x^2\, dx = \frac{x^3}{3}\Big|_0^2 = \frac{2^3}{3} = 2.7$

$\int_0^1 x^2\, dx = \frac{x^3}{3}\Big|_0^1 = \frac{1^3}{3} = \frac{1}{3}$

MATH WITH SPREADSHEETS

A spreadsheet program, such as Microsoft Excel and Apple Numbers, allows you to perform math calculations and manipulate numbers and data unlike any other tool. It saves time and allows you to see trends and easily forecast changes to come.

Most important, you can insert graphics and graphs into your presentations, which you can print and share with others.

Free spreadsheet software programs are available for download if your computer does not have one already installed.

This primer will show you some fundamental spreadsheet tips. The commands and functions use the Microsoft Excel style, which most spreadsheets follow (refer to your application's Help option for specific command information).

A SPREADSHEET PRIMER

■ Open a new spreadsheet file and you'll see columns and rows of boxes called cells, as shown in **FIGURE 1**. You can enter numbers and letters and data in each cell and refer to them in other cells.

For math functions, Excel recognizes the equal sign (=) as an operation to perform on numbers or data. If the contents of a cell do not start with an equal sign (=) the data may be considered a label (not numbers to perform calculations).

MICROSOFT OFFICE EXCEL 2010-MATH.xls

File Edit View Insert Format Tools Data Window Help

NA		▽ X ✓ f(x)		= SUM								
A	**B**	**C**	**D**	**E**	**F**	**G**	**H**	**I**	**J**	**K**	**L**	**M**
1												
2												

FIGURE 1 Spreadsheets allow you to easily visualize, present, and share your math.

FIGURE 2

=SUM (A1 + A2)

	A	B	C	D
1	10			
2	20			
3	=SUM (A1 + A2)			
4				
5				
6				

In cell A1, input 10.
In cell A2, input 20.
In cell A3, input formula: =SUM (A1 + A2)
Press Enter.

FIGURE 3

=SUM (A1 + A2)

	A	B	C	D
1	10			
2	20			
3	30			
4				
5				
6				

= SUM (A1 + A2)
Cell A3 displays the sum of
numbers in cells A1 and A2.

HERE'S A MORE PRACTICAL EXAMPLE: Input any numbers you prefer in the cells A1 and A2. In cell A3, type =SUM (A1 + A2) and press Enter, as shown in **FIGURE 2**.

Cell A3 will display the sum of the numbers in the cells above it. See **FIGURE 3**.

You can add, or sum, more numbers in columns and rows by using the : command like this: In cell A10 type: = SUM(A1:A9) and press Enter. This will sum all the numbers in a range from A1 to A9.

NOTE: Excel performs calculations from left to right and in some instances, you must place numbers in brackets to prevent errors. For instance, the command =1,000–100*5 could have an answer of 4,500 or 500. (1,000 minus 100 = 900 multiplied by 5 equals 4,500; or 1,000 minus [100 times 5], which is 500.)

You would enter this command in one of these two ways:

(1000 – 100) * 5   or   1000 – (100 * 5)

■ Excel offers many more built-in functions accessed from the Function menu shown in **FIGURE 4**. Statistical functions allow you to garner insights from data and determine the mean (average), median (middle), mode (most frequent), and standard deviation (span of a group) from a list of values. See **FIGURES 5, 6,** and **7**.

FIGURE 4

	A	B	C	D	E	F	Σ	
							SUM	
							AVERAGE	
1	10	20	30	40	50	60	COUNT	
							MAX	
2	20						MIN	
3	30						MORE FUNCTION	
4	40							
5								
6								

Autosum dropdown menu

FIGURE 5

	A	B	C	D	E	F	G	H	I	J	K	f_X
												AVERAGE
1	10											↓
2	20											FUNCTION ARGUMENT
3	30											↓
4	40											A1:A4
5	25											
6												

FIGURE 6

	A	B	C	D	E	F	G	H	I	J	K	f_X
												STEDEV
1	10											↓
2	20											FUNCTION ARGUMENT
3	30											↓
4	40											A1:A4
5	12.91											
6												

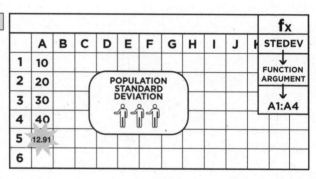

POPULATION STANDARD DEVIATION

FIGURE 7

	A	B	C	D	E	F	G	H	I	J	K	f_X
												MODE
1	10											↓
2	20											FUNCTION ARGUMENT
3	30											↓
4	40											A1:A4
5												
6												

FIGURE 8 | ⊠ **MICROSOFT EXCEL**

	A	B	C	D	E	F	G	H	I	J	K
1		MON	TUE	WED	THU						
2	CALEB	8	8	8	8						
3	BILL	7.5	8	6	3						
4	SUE	7.5	3	0	2						
5	JAKE	6	4	8	8						
6	AMY	4	7.5	8	8						

FIGURE 9 | ⊠ **MICROSOFT EXCEL**

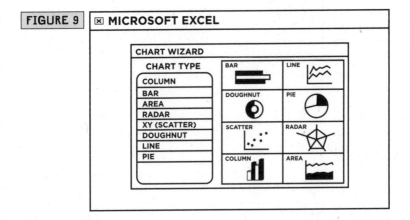

CHART WIZARD

CHART TYPE: COLUMN, BAR, AREA, RADAR, XY (SCATTER), DOUGHNUT, LINE, PIE — BAR, LINE, DOUGHNUT, PIE, SCATTER, RADAR, COLUMN, AREA

■ Additionally, you can insert graphics into your spreadsheets and add charts for a more appealing presentation. See **FIGURES 8** and **9**.

GOING FURTHER—SPREADSHEETS

- There are many keyboard and mouse shortcuts. Some allow you to select a group of cells, assign a command, or create a formula for you!

- After working with the basic commands and functions, study your spreadsheet manual and tutorials to discover more functions, including text and data functions, date and time, math and trigonometry, and statistical analysis:

  ▸ Financial Formulas—Interest Calculation, Future Value, Loan Payments

  ▸ Calculators, What-If Analysis, Regression Analysis

ADDITIONAL MATH DEMO DESIGNS

This book has presented physical projects as a way to make math memorable, fun, and easy to share. You can go further by making more DIY projects to help inspire math students and keep their attention.

The following designs are just a few examples you can make with everyday things:

■ Make *giant* versions of the Sneaky Math Quizzers shown in this book and mount them on a wall.

■ Repetitive viewing aids learning, especially with unusual symbols and formulas. Make large, dynamic Sneaky Math symbol posters and mount them on your walls. See **FIGURE 1**.

■ Take a break from formal study to create and play math strategy games such as "Wheel of Function," as seen in **FIGURE 2**.

Crafty Designs for the Classroom and Home

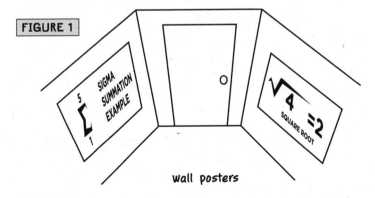

FIGURE 1

wall posters

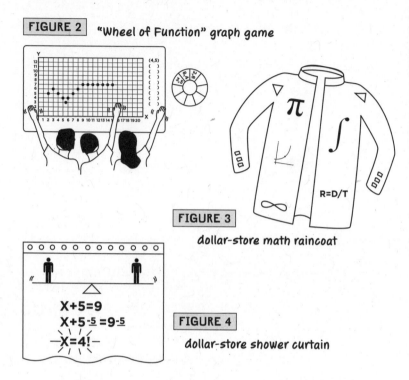

FIGURE 2 "Wheel of Function" graph game

FIGURE 3

dollar-store math raincoat

FIGURE 4

dollar-store shower curtain

- Decorate inexpensive vinyl ponchos and shower curtains with educational math designs as shown in **FIGURES 3** and **4**.

- Discarded cookie or cracker boxes can be crafted into sliding Algebra Function Demo devices. See **FIGURES 5** and **6**.

- Don't leave your math cards unseen in a drawer. Make a math card caddy: turn a cereal box inside out, refold it, tape it together, and cut slits so you can slide the math cards on the front and back. See **FIGURE 7**.

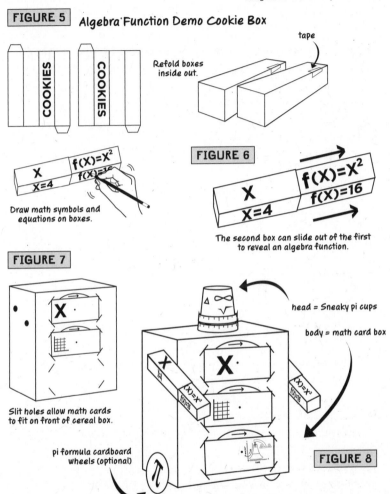

FIGURE 5 Algebra Function Demo Cookie Box

COOKIES COOKIES

Refold boxes inside out.

tape

Draw math symbols and equations on boxes.

X
X=4
$f(X)=X^2$
$f(X)=16$

FIGURE 6

X
X=4
$f(X)=X^2$
$f(X)=16$

The second box can slide out of the first to reveal an algebra function.

FIGURE 7

X

Slit holes allow math cards to fit on front of cereal box.

pi formula cardboard wheels (optional)

head = Sneaky pi cups

body = math card box

X

$f(X)=X^2$

FIGURE 8

- If you've crafted Sneaky Pi Cups, built sliding Algebra Function Demo devices, and produced a Sneaky Math card caddy, put them together with Velcro strips (or tape) to make a mobile Sneaky Mathbot, as shown in **FIGURE 8**.

MORE CREATIVE MATH DESIGNS AND CHALLENGES USING EVERYDAY THINGS

Make Sneaky Math book covers so you'll learn day-by-day when you carry your textbooks. Draw images and formulas from this book's Glossary pages on large paper as shown in **FIGURE 1**.

Then fold over the top and bottom sections and slip the ends over the book covers. See **FIGURE 2**.

Over time you'll see and learn math principles when you look at your schoolbooks. See **FIGURE 3**.

Test yourself and others with a measurement quiz. See who can calculate the dimensions of currency, **FIGURE 4**, and coins, **FIGURE 5**, using the Pythagorean theorem and the formula for circumference.

Then find the circumference of common objects like soda cans and batteries, as shown in **FIGURE 6**.

Don't stop there. Celebrate "Pi Day" every March 14 (3/14). (Check out the Sneaky Pi Cups on p. 79.) Think of more ways to make math visual and physical. Repurpose toys, radio-controlled vehicles, Lego-like craft kits, and party and board games such as Twister, Charades, and Scrabble into fun math activities!

Discover how math is used in physics from books including *Mathletics*, *Space Mathematics*, and *The Physics of Superheroes*—see the Recommended Books section on p. 167.

For additional crafty concepts, see *Sneaky Uses for Everyday Things* and other books in the series at www.sneakyuses.com.

FIGURE 1

π pi= 3.1415

Draw or copy glossary entries onto 11 x 17 paper

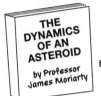

THE DYNAMICS OF AN ASTEROID

by Professor James Moriarty

book

FIGURE 2

book

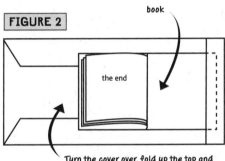

the end

Turn the cover over, fold up the top and bottom, and slide the ends over the book's front and back covers.

FIGURE 3

π pi= 3.1415

Sneaky Math cover installed on book.

FIGURE 4 Sneaky Math Challenge

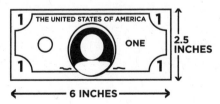

THE UNITED STATES OF AMERICA

1 1

ONE

1 1

2.5 INCHES

← 6 INCHES →

US dollar bills are 6 inches by 2.5 inches. Using the Pythagorean theorem, what is the diagonal measurement? Clue: $A^2 + B^2 = C^2$

FIGURE 5 Sneaky Math Challenge

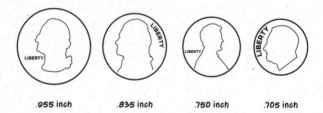

.955 inch   .835 inch   .750 inch   .705 inch

Given the diameters of the coins above, find their circumferences.
Clue: C = πD
circumference = π x diameter

FIGURE 6 Sneaky Math Challenge

12 oz. soda can   D battery   C battery   AA battery
2.5-inch diameter   1.3-inch diameter   1.032-inch   .53-inch diameter
                                            diameter

Given the diameters of the soda can and batteries,
find their circumferences.
Clue: C = πD
Circumference = pi x diameter

SNEAKY MATH
REFERENCE MATERIAL

MODERN MATH NOTABLES

Bill James
By analyzing the game in his *Baseball Abstracts* in the '70s and '80s, Bill James altered preconceived ideas about American baseball.

Since 1977, James has written more than two dozen books devoted to baseball history and statistics. His approach, which he termed *sabermetrics* in reference to the Society for American Baseball Research (SABR), scientifically analyzes and studies baseball, often through the use of statistical data, in an attempt to determine why teams win and lose. He is featured in the book *Moneyball* and in the film version of the book.

billjamesonline.com

Salman Amin "Sal" Khan
Former hedge fund analyst Salman Khan is founder of the Khan Academy, a free online education platform and nonprofit organization. From a small office in his home, Khan has produced more than 4,000 video lessons teaching a wide spectrum of academic subjects, mainly focusing on mathematics and the sciences. In 2012, *Time* named Salman Khan in its annual list of the 100 most influential people in the world.

khanacademy.org

Danica McKellar

Danica McKellar is an American actress, film director, academic, book author, and education advocate. She is best known for her role as Winnie Cooper in the television show *The Wonder Years*, and later as the *New York Times* best-selling author of four nonfiction books: *Math Doesn't Suck*, *Kiss My Math*, *Hot X: Algebra Exposed*, and *Girls Get Curves: Geometry Takes Shape*. McKellar's books encourage middle- and high-school girls to have confidence and succeed in mathematics.

danicamckellar.com

Top Secret Rosies: The Female Computers of WWII

On Sunday, December 7, 1941, the Japanese bombed Pearl Harbor and changed these young women's lives forever. With Pearl Harbor suddenly drawing the US into World War II, the Army launched a frantic national search for women mathematicians.

In 1942, when computers were human and women were underestimated, a group of female mathematicians helped win a war and usher in the modern computer age. Sixty-five years later their story has finally been told.

Code-named Top Secret Rosies, these female mathematicians and scientists were secretly recruited to do ballistics research and crack codes during WWII. (They were called "female computers," back in the days when "computer" meant "one who computes.")

topsecretrosies.com

Glen Whitney

Glen Whitney, a hedge fund manager turned mathematics advocate, opened the Museum of Mathematics, the only museum of its kind, in New York City. Whitney serves as a member of the Board of Education for the Three Village School District on Long Island, New York, where he is a strong advocate for the value of mathematics in education.

momath.org

RECOMMENDED BOOKS

For Young Readers

The Book of Perfectly Perilous Math: 24 Death-Defying Challenges for Young Mathematicians by Sean Connolly (Workman Publishing Company, 2012)

Help Your Kids with Math: A Visual Problem Solver for Kids and Parents by Barry Lewis (DK Publishing, 2010)

Math and Popular Culture

The Joy of X: A Guided Tour of Math, from One to Infinity by Steven Strogatz (Mariner Books, 2013)

Math Goes to the Movies by Burkard Polster and Marty Ross (Johns Hopkins University Press, 2012)

The Simpsons and Their Mathematical Secrets by Simon Singh (Bloomsbury USA, 2013)

Math and Physics

Mathletics: A Scientist Explains 100 Amazing Things About the World of Sports by John D. Barrow (W.W. Norton & Company, 2012)

Space Mathematics: Math Problems Based on Space Science by Bernice Kastner (Dover Publications, 2012)

The Physics of Superheroes by James Kakalios (Gotham, 2006)

Don't Try This at Home!: The Physics of Holiday Movies by Adam Weiner (Kaplan Publishing Company, 2007)

Comprehensive Titles

Math Doesn't Suck, 2008; *Hot X: Algebra Exposed!*, 2011; *Girls Get Curves: Geometry Takes Shape*, 2013 by Danica McKellar (Plume)

Calculus for Cats by Kenn Amdahl and Jim Loats (Clearwater Publishing, 2001)

The Calculus Diaries: How Math Can Help You Lose Weight, Win in Vegas, and Survive a Zombie Apocalypse by Jennifer Ouellette (Penguin Books, 2010)

Calculus for Dummies, 2003, *Calculus Workbook for Dummies*, 2005, by Mark Ryan (For Dummies)

Information Wheel Designs Similar to the Sneaky Math Cards

Reinventing the Wheel by Jessica Helfand (Princeton Architectural Press, 2006)

RECOMMENDED WEBSITES

- www.khanacademy.org

- www.mathisfun.com

- www.coolmath.com

- www.easycalculation.com

- www.momath.org
  (Museum of Mathematics)

- www.math.com
  (for online scientific
  calculators)

- www.docs.google.com
  (for online spreadsheet)

- spacemath.gsfc.nasa.gov
  (math and calculus lessons
  and more)

- www.algebra.com

- www.piday.org
  (Pi Day website)

- www.nctm.org
  (National Council of
  Teachers of Mathematics)

- www.sneakyuses.com

- www.sneakymath.com

GLOSSARY

$\approx$ **Approximately**
Approximately equal to

$\circ$ **Degrees**
Indication of temperature amount or a degree of an arc.

Δ **Delta**
"Delta" means "change in."

e **Euler**
The small letter e stands for the Euler number of 2.7118281828459045. It is used with interest payment plans and other scientific calculations.

$n!$ **Factoral**
The sign $n!$ is a space-saving symbol.
EXAMPLE: $4! = 4 \times 3 \times 2 \times 1$

$f(x)=$ **Function**
A function is the relationship between two variables. A function is a predefined formula using the f(x) notation. It means "function of x."

$\neq$
$\geq$
$\leq$ **Inequality**
Not equal to
Greater than or equal to
Less than or equal to.

∞ **Infinity**
"Infinity" means "to go on forever" or towards infinity
EXAMPLE: Infinitely small.

$\int$ **Integral**
The integral sign is used for integration of, or
combining together, items.

$\cap$ **Intersect**
The common data or terms that two or more number
sets share. **EXAMPLE:** (1,2,3,4,5) (3,4) Intersect = (3,4)

$\lim h \to 0$ **Limit**
The value that a function approaches at a given X

$x \bullet ()$ **Multiplication sign**
Multiplication examples:
$2 \times 3 = 6$ $2 \bullet 3 = 6$ $2(3) = 6$

% **Percent**
A number or ratio that is a fraction of 100

π **Pi**
The ratio of the diameter of a circle to the
circumference is called pi. Pi is a number
that never ends or repeats and begins with
3.14159265358979323.

 **Radical**
The radical symbol is used for square root calculations.
The number next to the radical is called the radicand.

: **Range/Ratio**
Indicates a ratio of two terms

T:{(2,6),(4,1)} **Relation**
Ordered pair of specific input numbers and their output results

Σ **Summation**
The summation, sum, or sigma symbol is used for adding together numbers in a range. Numbers above and below the sum symbol designate how many times to add the numbers in increments to a variable number. The first time, the number 1 is added to the variable 1. Then 2 is added and so forth.

θ **Theta**
The theta symbol represents an angle.

$\cup$ **Union**
The common data or terms that two or more number sets share. (1,2,3,4,5) (3,4)   union = (1,2,3,4,5)

X **Variable**
A variable is a placeholder for an amount that you do not know. A variable, like the letter X, can represent time, distance, money, or people, and allows you to make equations and time-saving formulas.

⌒ **Vector**
Vector denotes direction and magnitude.

MATH CONVERSION CHART

Imperial	Metric
1 inch	2.54 centimeters
1 foot	0.3048 meters
1 yard	0.9144 meters
1 mile	1.6093 kilometers
1 nautical mile	1.853 kilometers
1 ounce	28.349 grams
1 pound	0.453 kilograms
1 stone	6.3503 kilograms
1 hundredweight	50.802 kilograms
1 imperial ton	1.016 metric tons
1 pint	0.473 liters
1 gallon	3.785 liters

TEMPERATURE CONVERSION FORMULAS

Fahrenheit to Celsius	$C=(F-32)\times5\div9$
Celsius to Fahrenheit	$F=(C\times9\div5)+32$
Celsius to Kelvin	$K=C+273$
Kelvin to Celsius	$C=K-273$

FRACTION TO DECIMAL CONVERSION TABLES

fraction = decimal			
$\frac{1}{1} = 1$			
$\frac{1}{2} = 0.5$			
$\frac{1}{3} = 0.\underline{3}$	$\frac{2}{3} = 0.\underline{6}$		
$\frac{1}{4} = 0.25$	$\frac{3}{4} = 0.75$		
$\frac{1}{5} = 0.2$	$\frac{2}{5} = 0.4$	$\frac{3}{5} = 0.6$	$\frac{4}{5} = 0.8$
$\frac{1}{6} = 0.1\underline{6}$	$\frac{5}{6} = 0.8\underline{3}$		
$\frac{1}{7} = 0.142857$	$\frac{2}{7} = 0.285714$	$\frac{3}{7} = 0.428571$	$\frac{4}{7} = 0.571428$
	$\frac{5}{7} = 0.714285$	$\frac{6}{7} = 0.857142$	
$\frac{1}{8} = 0.125$	$\frac{3}{8} = 0.375$	$\frac{5}{8} = 0.625$	$\frac{7}{8} = 0.875$
$\frac{1}{9} = 0.\underline{1}$	$\frac{2}{9} = 0.\underline{2}$	$\frac{4}{9} = 0.\underline{4}$	$\frac{5}{9} = 0.\underline{5}$
	$\frac{7}{9} = 0.\underline{7}$	$\frac{8}{9} = 0.\underline{8}$	
$\frac{1}{10} = 0.1$	$\frac{3}{10} = 0.3$	$\frac{7}{10} = 0.7$	$\frac{9}{10} = 0.9$
$\frac{1}{11} = 0.\underline{09}$	$\frac{2}{11} = 0.\underline{18}$	$\frac{3}{11} = 0.\underline{27}$	$\frac{4}{11} = 0.\underline{36}$
	$\frac{5}{11} = 0.\underline{45}$	$\frac{6}{11} = 0.\underline{54}$	$\frac{7}{11} = 0.\underline{63}$
	$\frac{8}{11} = 0.\underline{72}$	$\frac{9}{11} = 0.\underline{81}$	$\frac{10}{11} = 0.\underline{90}$
$\frac{1}{12} = 0.08\underline{3}$	$\frac{5}{12} = 0.41\underline{6}$	$\frac{7}{12} = 0.58\underline{3}$	$\frac{11}{12} = 0.91\underline{6}$
$\frac{1}{16} = 0.0625$	$\frac{3}{16} = 0.1875$	$\frac{5}{16} = 0.3125$	$\frac{7}{16} = 0.4375$
	$\frac{11}{16} = 0.6875$	$\frac{13}{16} = 0.8125$	$\frac{15}{16} = 0.9375$
$\frac{1}{32} = 0.03125$	$\frac{3}{32} = 0.09375$	$\frac{5}{32} = 0.15625$	$\frac{7}{32} = 0.21875$
	$\frac{9}{32} = 0.28125$	$\frac{11}{32} = 0.34375$	$\frac{13}{32} = 0.40625$
	$\frac{15}{32} = 0.46875$	$\frac{17}{32} = 0.53125$	$\frac{19}{32} = 0.59375$
	$\frac{21}{32} = 0.65625$	$\frac{23}{32} = 0.71875$	$\frac{25}{32} = 0.78125$
	$\frac{27}{32} = 0.84375$	$\frac{29}{32} = 0.90625$	$\frac{31}{32} = 0.96875$

NOTE-underlined numbers indicates those numbers are repeatd.

POWER TABLE

Number	Squared	Cubed	Fourth	Fifth	Sixth
1	1	1	1	1	1
2	4	8	16	32	64
3	9	27	81	243	729
4	16	64	256	1,024	4,096
5	25	125	625	3,125	15,625
6	36	216	1,296	7,776	46,656
7	49	343	2,401	16,807	117,649
8	64	512	4,096	32,768	262,144
9	81	729	6,561	59,049	531,441
10	100	1,000	10,000	100,000	1,000,000
11	121	1,331	14,641	161,051	1,771,561
12	144	1,728	20,736	248,832	2,985,984
13	169	2,197	28,561	371,293	4,826,809
14	196	2,744	38,416	537,824	7,529,536
15	225	3,375	50,625	759,375	11,390,625
16	256	4,096	65,536	1,048,576	16,777,216
17	289	4,913	83,521	1,419,857	24,137,569
18	324	5,832	104,976	1,889,568	34,012,224
19	361	6,859	130,321	2,476,099	47,045,881
20	400	8,000	160,000	3,200,000	64,000,000

Number	Squared	Cubed	Fourth	Fifth	Sixth
21	441	9,261	194,481	4,084,101	85,766,121
22	484	10,648	234,256	5,153,632	113,379,904
23	529	12,167	279,841	6,436,343	148,035,889
24	576	13,824	331,776	7,962,624	191,102,976
25	625	15,625	390,625	9,765,625	244,140,625
26	676	17,576	456,976	11,881,376	308,915,776
27	729	19,683	531,441	14,348,907	387,420,489
28	784	21,952	614,656	17,210,368	481,890,304
29	841	24,389	707,281	20,511,149	594,823,321
30	900	27,000	810,000	24,300,000	729,000,000

TANGENT TABLE

Degree Angle	Tangent	Degree Angle	Tangent
0°	0	16°	0.28675
1°	0.01746	17°	0.30573
2°	0.03492	18°	0.32492
3°	0.05241	19°	0.34433
4°	0.06993	20°	0.36397
5°	0.08749	21°	0.38386
6°	0.10510	22°	0.40403
7°	0.12278	23°	0.42447
8°	0.14054	24°	0.44523
9°	0.15838	25°	0.46631
10°	0.17633	26°	0.48773
11°	0.19438	27°	0.50953
12°	0.21256	28°	0.53171
13°	0.23087	29°	0.55431
14°	0.24933	30°	0.57735
15°	0.26795		

Degree Angle	Tangent	Degree Angle	Tangent
31°	0.60086	46°	1.03553
32°	0.62487	47°	1.07237
33°	0.64941	48°	1.11061
34°	0.67451	49°	1.15037
35°	0.70021	50°	1.19175
36°	0.72654	51°	1.23490
37°	0.75355	52°	1.27994
38°	0.78129	53°	1.32704
39°	0.80978	54°	1.37638
40°	0.83910	55°	1.42815
41°	0.86929	56°	1.48256
42°	0.90040	57°	1.53986
43°	0.93252	58°	1.60033
44°	0.96569	59°	1.66428
45°	1	60°	1.73205

Degree Angle	Tangent	Degree Angle	Tangent
61°	1.80405	76°	4.01078
62°	1.88073	77°	4.33148
63°	1.96261	78°	4.70463
64°	2.05030	79°	5.14455
65°	2.14451	80°	5.67128
66°	2.24604	81°	6.31375
67°	2.35585	82°	7.11537
68°	2.47509	83°	8.14435
69°	2.60509	84°	9.51436
70°	2.74748	85°	11.43005
71°	2.90421	86°	14.30067
72°	3.07768	87°	19.08114
73°	3.27085	88°	28.63625
74°	3.48741	89°	57.28996
75°	3.73205	90°	Undefined

(90-degrees is undefined because it is a vertical line with no angle)

NOTES

NOTES

NOTES

NOTES

SNEAKY MATH
A Graphic Primer with Projects

Andrews McMeel Publishing, LLC
an Andrews McMeel Universal company
1130 Walnut Street, Kansas City, Missouri 64106

www.andrewsmcmeel.com

14 15 16 17 18 RR2 10 9 8 7 6 5 4 3 2 1

ISBN: 978-1-4494-4520-1

Library of Congress Control Number: 2014935898

ATTENTION: SCHOOLS AND BUSINESSES
Andrews McMeel books are available at quantity discounts
with bulk purchase for educational, business, or sales
promotional use. For information, please e-mail the
Andrews McMeel Publishing Special Sales Department:
specialsales@amuniversal.com.